T0222157

Lying by Approximation

The Truth about Finite Element Analysis

Synthesis Lectures on Engineering

Each book in the series is written by a well known expert in the field. Most titles cover subjects such as professional development, education, and study skills, as well as basic introductory undergraduate material and other topics appropriate for a broader and less technical audience. In addition, the series includes several titles written on very specific topics not covered elsewhere in the Synthesis Digital Library.

Lying by Approximation: The Truth about Finite Element Analysis
Vincent C. Prantil, Christopher Papadopoulos, and Paul D. Gessler
2013

The Engineering Design Challenge: A Creative Process
Charles W. Dolan
2013

The Making of Green Engineers: Sustainable Development and the Hybrid Imagination
Andrew Jamison
2013

Crafting Your Research Future: A Guide to Successful Master's and Ph.D. Degrees in Science & Engineering
Charles X. Ling and Qiang Yang
2012

Fundamentals of Engineering Economics and Decision Analysis
David L. Whitman and Ronald E. Terry
2012

A Little Book on Teaching: A Beginner's Guide for Educators of Engineering and Applied Science
Steven F. Barrett
2012

Engineering Thermodynamics and 21st Century Energy Problems: A Textbook Companion for Student Engagement
Donna Riley
2011

MATLAB for Engineering and the Life Sciences
Joseph V. Tranquillo
2011

Systems Engineering: Building Successful Systems
Howard Eisner
2011

Fin Shape Thermal Optimization Using Bejan's Constructal Theory
Giulio Lorenzini, Simone Moretti, and Alessandra Conti
2011

Geometric Programming for Design and Cost Optimization (with illustrative case study problems and solutions), Second Edition
Robert C. Creese
2010

Survive and Thrive: A Guide for Untenured Faculty
Wendy C. Crone
2010

Geometric Programming for Design and Cost Optimization (with Illustrative Case Study Problems and Solutions)
Robert C. Creese
2009

Style and Ethics of Communication in Science and Engineering
Jay D. Humphrey and Jeffrey W. Holmes
2008

Introduction to Engineering: A Starter's Guide with Hands-On Analog Multimedia Explorations
Lina J. Karam and Naji Mounsef
2008

Introduction to Engineering: A Starter's Guide with Hands-On Digital Multimedia and Robotics Explorations
Lina J. Karam and Naji Mounsef
2008

CAD/CAM of Sculptured Surfaces on Multi-Axis NC Machine: The DG/K-Based Approach
Stephen P. Radzevich
2008

Lying by Approximation: The Truth about Finite Element Analysis
Vincent C. Prantil, Christopher Papadopoulos, and Paul D. Gessler

ISBN:978-3-031-79362-2   paperback
ISBN:978-3-031-79363-9   ebook

DOI 10.1007/978-3-031-79363-9

A Publication in the Springer series
SYNTHESIS LECTURES ON ENGINEERING

Lecture #23
Series ISSN
Synthesis Lectures on Engineering
Print 1939-5221   Electronic 1939-523X

Lying by Approximation

The Truth about Finite Element Analysis

Vincent C. Prantil
Milwaukee School of Engineering

Christopher Papadopoulos
University of Puerto Rico Mayagüez

Paul D. Gessler
Graduate Student, Marquette University

SYNTHESIS LECTURES ON ENGINEERING #23

ABSTRACT

In teaching an introduction to the finite element method at the undergraduate level, a prudent mix of theory and applications is often sought. In many cases, analysts use the finite element method to perform parametric studies on potential designs to size parts, weed out less desirable design scenarios, and predict system behavior under load. In this book, we discuss common pitfalls encountered by many finite element analysts, in particular, students encountering the method for the first time. We present a variety of simple problems in axial, bending, torsion, and shear loading that combine the students' knowledge of theoretical mechanics, numerical methods, and approximations particular to the finite element method itself. We also present case studies in which analyses are coupled with experiments to emphasize validation, illustrate where interpretations of numerical results can be misleading, and what can be done to allay such tendencies. Challenges in presenting the necessary mix of theory and applications in a typical undergraduate course are discussed. We also discuss a list of tips and rules of thumb for applying the method in practice.

KEYWORDS

finite element method, finite element analysis, numerical methods, computational analysis, engineering mechanics, mathematical modeling, modeling approximation

Contents

Preface

WHAT THIS BOOK IS INTENDED TO BE

In undergraduate engineering curricula, a first course in finite element analysis (FEA) is routinely required, but is often not taken until after the second year of study. Such a class typically includes:

1. an overview of the procedural aspects of the method;

2. a derivation of the mathematical theory for a variety of relatively simple one- and two-dimensional element formulations;

3. practicing the finite element procedure by hand on select simple problems; and

4. employing the finite element method using some commercial software package as practiced by engineers in industry.

Students are increasingly expected to apply this knowledge in other settings, particularly in the context of their senior capstone design projects. However, students routinely commit a variety of errors in applying FEA. In particular, they lack the maturity to make appropriate modeling decisions and interpretations of their results. This, in turn, inhibits them from using FEA to make sound judgements in their projects. Indeed, the twin abilities to conduct accurate analyses *and* to make informed judgements lie at the heart of what it means to be a competent professional engineer.

Many instructors are aware of this circumstance and recognize the need to coach their students to perform FEA with greater maturity, but they are often mired in teaching strictly according to the treatment of standard textbooks which emphasize underlying derivation and theory. Indeed, there is a need for such deep, rigorous, and detailed study, but not at the expense of learning mature habits. Many professors therefore develop means to *teach around the text* by providing additional explanations, insights, approaches, and probing questions.

Our intent here is to provide just such an alternative resource for professors and instructors of undergraduates who are looking for a fresh and novel approach to teaching FEA that prioritizes the development of practical skills and good habits. Using material compiled from existing course notes and exercises already in use by the authors and their colleagues, we lay a path through the forest of details that an undergraduate or other novice can follow to discover the habits and secrets of a seasoned user. We surmise that in this book already lie many ideas that match what many instructors already intuitively understand and convey as they, on their own, *teach around the text*.

In laying this path, we deliberately employ an approach to emphasize and exploit the natural ties between classical Mechanics of Materials (MoM) and FEA, and which is motivated, in part, by the philosophy articulated in Papadopoulos et al. [2011]. Of course, equally deep ties exist between elasticity theory and FEA, but as our focus is developing expertise of undergraduates, we appeal primarily to the ties between FEA and MoM.

In this approach, we provide examples in which FEA can be used to confirm results of hand calculation, closed form solutions, or standard tables—and vice versa—helping students to build confidence in all. The book then explores more advanced user habits such as formulating expectations, making estimates, and performing benchmark calculations. Broadly speaking, this book responds to the growing call to include simulation as a basic engineering competency, and will help to promote the development of a culture of using simulation in the undergraduate engineering curriculum.

As such, we envision this book being used as a companion to a traditional textbook in an upper-level undergraduate FEA course and also as an instructional guide for practice in other courses in which FEA is applied, including courses as early as freshman design and introductory mechanics. Even at these early stages, instructors can judiciously draw from the book to plant the seeds of good habits in their students. This book is written in language that is immediately transparent to instructors and accessible to students who have completed a basic course in MoM. Terminologies that might be advanced to the novice user are italicized and explained in the context of their use.

PEDAGOGICAL APPROACH

The pedagogical strategy of this book is based in the educational theory of *constructivism* and related research in misconceptions. The essence of constructivist philosophy to which we appeal here is rooted in the work of cognitive psychologist Jerome Bruner, and is succinctly described by Montfort et al. [2009]: "learning [is] a complex process in which learners are constantly readjusting their existing knowledge and, more importantly, the relationships between the things that they know." Further, this readjustment process requires that the learner not just passively receive information, but actively enter into the "discovery of regularities of previously unrecognized relations and similarities between ideas, with a resulting sense of self-confidence in one's abilities" [Bruner, 1960].

One way to involve students in the processes of readjusting and discovering knowledge is by anticipating their misconceptions and providing exercises and activities that force them to reevaluate their original assumptions and conceptions. For at least three decades, science and engineering educators have realized the importance of identifying and addressing misconceptions, suggesting that educators should directly address misconceptions by some combination of early intervention and an infusion of activities that force students to face the misconceptions head-on [Hake, 1998, McDermott, 1984, Montfort et al., 2009, Papadopoulos, 2008, Streveler et al., 2008]. Broadly speaking, "active learning," "problem based learning," "inquiry based learning," and "student centered learning" approaches aim to accomplish this. Ken Bain, in his book, *What the Best College Teachers Do*, champions this view:

> Some of the best teachers want to create an *expectation failure*, a situation in which existing mental models lead to faulty expectations. They attempt to place students in situations where their mental models will not work. They listen to student conceptions before challenging them. They introduced problems, often case studies of what could go wrong, and engaged the students in grappling with the issues those examples raised [Bain, 2004].

Physics educator Lillian McDermott further adds that "students need to participate in the process of constructing qualitative models and applying these models to predict and explain real-world phenomena" [McDermott, 2001].

It is important to observe that this type of instruction requires a high degree of interaction and feedback on the part of the teacher and a correspondingly high degree of self-inquiry on the part of the learner. In this environment, teachers need to allow students to test ideas, and lend support in tweaking those ideas into a more correct model of how things happen, and students must eagerly participate in this process of discovery.

In the spirit of those instructors who have successfully accomplished this, we seek to provide students with the support they need to cognitively rewire. Indeed, many of the examples and exercises are deliberately designed to confront readers with expectation failures and to provide them ample opportunity to develop models that appropriately match reality, but which also require instructors to intervene as supportive mentors. With this approach, novices and students will develop the good habits required of experienced users.

In the particular case of FEA, many of the common pitfalls repeatedly encountered by analysts are rooted in a mixture of inadequacies in their understanding of MoM theory, modeling, and the useful approximations particular to FEA, as well as their inability to integrate these areas of knowledge. To address these matters, we aim to strike a prudent balance between theory and practical application. We suggest that this is best accomplished by prescribing a *minimal requisite skill set*, rooted in mastery of MoM, upon which the modeling decisions required in the finite element method are based. This mastery of the most rudimentary underlying theory helps students make fewer of the errors in judgement when validating their numerical simulations.

Ultimately, our emphasis is to provide an instructional approach that is amenable to a practicing engineer rather than a mathematician. We attempt to cultivate the habit of care that is necessary to perform good quality engineering analysis. When answering the question "What is a university for?," New York Times columnist David Brooks wrote:

> [to obtain] technical knowledge and practical knowledge. Technical knowledge is formulas…that can be captured in lectures. Practical knowledge is not about what you do, but how you do it. It can not be taught or memorized, only imparted and absorbed. It is not reducible to rules; it only exists in practice [Brooks, 2013].

In view of this attitude toward practice, we provide guidance for using pre-programmed software. Guidance is offered for both commercial software and academically developed finite element codes via the online video tutorials found at the wiki site SimCafe (`https://confluence.cornell.edu/display/SIMULATION/Home`) [Bhaskaran, 2012]. The NSF-sponsored project team at Cornell University [Bhaskaran and Dimiduk, 2010] has graciously supplied ANSYS tutorials for the collection of illustrative case studies presented here. All tutorials for this book can be found at `https://confluence.cornell.edu/display/SIMULATION/Prantil+et+al`.

In summary, we write this book for student and faculty colleagues who are willing to undertake

1. iterative learning in a supportive environment in which students are unafraid to make errors, confront misconceptions, and revisit problems, and in which instructors are present to provide support "when things go wrong;"

2. a strong navigational approach that is orderly and progressive, but not necessarily "top down;" and

3. an approach in which MoM theory and FEA are intimately entwined.

WHAT THIS BOOK IS NOT INTENDED TO BE

Most current textbook treatments of the mathematical theory of finite elements draw on variational calculus and linear algebra. As suggested previously, we intend this book to serve as a supplement for more advanced undergraduates and as a resource to inform teaching of earlier stage students. Our focus is not on treatment of the mathematical rigor and underpinnings of the finite element method, but rather a guide to good practice. Therefore, this book is not intended to be a reference or text on the formulation, theory, or mathematical underpinnings of the finite element method. There are many excellent treatments outlining the method [Cook et al., 2002, Kim and Sankar, 2009, Logan, 2001, Thompson, 2004, Zienkiewicz and Taylor, 2005, Zienkiewicz et al., 2005]. Any one of these would be sufficient for an introductory course in an undergraduate mechanical engineering curriculum.

This book is also not intended to be a tutorial guide for applying the method or a step-by-step user's guide to a particular commercial software package, e. g., Kurowski [2013], Lawrence [2012], Lee [2012]. We assume that the instructor using this book is already providing such tutorial instruction or that the reader already has a working knowledge of such. We emphasize, however, that we do provide online video tutorials at the SimCafe wiki site, which include further user guidance and suggested follow-up exercises. *We encourage the student or novice reader to open a tutorial or start an FEA session from scratch and directly attempt the exercises and examples that are provided in both the tutorials and the book.*

OUTLINE OF BOOK

Chapter 1 addresses why humans tend to have an optimism bias in which they think they are correct in more situations than they really are [Conly, 2013]. Digital technology has most likely added to this bias. We review a published list of ten common mistakes made in FEA practice, and we argue that avoidance of these errors begins with the user adopting an attitude of *skepticism* of numerical results until they have been validated. Most analysts agree this is best done by comparison with relevant theory and experimental data. To apply theory, one must be fluent in the very basic mechanics relationships.

In Chapter 2 we summarize essential topics from Mechanics of Materials and provide corresponding examples that can be solved using simple, well-known relationships based on one-dimensional modeling assumptions. While these problems do not require use of FEA, they are excellent for offering a first exposure to FEA in which the user can quickly build confidence in the method. Moreover, the theory underlying these examples forms the basis for the "minimal requisite skill set" mentioned previously. With this in hand, the user can begin the crucial task of understanding how to interpret FEA results by comparison with a trusted theory. This small set of topics is remarkably useful due to the great number of situations in which they serve as good models for practical situations.

However, as problems become more detailed and complex, the applicability of these elementary relations diminishes. Here, a more complex multi-dimensional theory of elasticity may be required but FEA can still be used to obtain reasonable approximate solutions, and basic principles from Mechanics of Materials can still be applied to interpret results, albeit with caution. Therefore, in Chapter 3, we illustrate several examples of problems whose analytical solutions (where tractable) are more involved, and where FEA is eminently useful, although still relatively straightforward.

Chapter 4 gets at the core of the list of the common mistakes made when pre-processing the finite element model. Mistakes that plague many finite element analysts involve relatively simple errors in input that seem intuitively correct, but which have strong adverse consequences for numerical predictions of displacements and stresses.

Finally, in Chapter 5, we present a list of prudent practices as well as pitfalls to avoid in order to achieve meaningful results and to make validation of one's results a less onerous task. This chapter can serve as an excellent reference as the reader begins to venture in his or her own practice.

Vincent C. Prantil, Christopher Papadopoulos, and Paul D. Gessler
August 2013

Acknowledgments

We gratefully acknowledge Rajesh Bhaskaran, Director of the Swanson Engineering Simulation Program at Cornell University, and Robert McBride for providing the ANSYS finite element tutorials at the wiki site SimCafe (`https://confluence.cornell.edu/display/SIMULATION/Home`). We also gratefully acknowledge funding provided to Cornell University under Award 0942706 from the National Science Foundation (NSF) for implementation of the SimCafe wiki interface.

We greatly appreciate the contributions of Jim Papadopoulos of Northeastern University, who has long been a passionate advocate for introducing FEA practice throughout the undergraduate engineering curriculum. His vision, keen insights, and collaboration on a previous article inspired ideas used in this book. We are grateful to him and also to Habib Tabatabai from the University of Wisconsin–Milwaukee for providing a review of an earlier draft of this work which helped us to polish and refine many details. We also acknowledge William E. Howard of East Carolina University. His collaboration on several previous articles linking simulations and experiments formed the basis of several examples in this book.

We also express our sincere gratitude to our colleagues Genock Portela Gauthier and Aidsa Santiago Roman of the University of Puerto Rico, Mayagüez. They are collaborators on a related project sponsored by the National Science Foundation under Award 1044886 that is developing new simulation tools for mechanics courses. Some of the modules developed in this NSF project and related understandings of engineering pedagogy appear in this book.

We further acknowledge personal communications and discussion that took place at the 9th U.S. National Congress on Computational Mechanics held in San Francisco in July 2007. Professors Jat du Toit, Mike Gosz, and Göran Sandberg organized the first mini-symposium on the teaching of finite element methods to undergraduates at which some of the first ideas for this supplementary text were discussed and took preliminary form. We are grateful to the mini-symposium for both fostering an international debate and proffering fruitful discussions leading to this work.

Finally, we acknowledge chapter heading character designs illustrated by Tim Decker, Senior Lecturer at the Peck School of the Arts at the University of Wisconsin–Milwaukee and Milwaukee Area Technical College. Prior to teaching animation in Milwaukee, Tim was the layout artist and animator for the award-winning television series "The Simpsons!," and animation supervisor for Disney Interactive. He has also appeared as a guest artist in animation and cartooning for PBS. We are grateful for Tim's imaginative characterization of numerical analysts practicing the fine art of approximation.

From Vincent C. Prantil: I wish to dedicate this book to my wife, Laurna, and my children, Carmen and Lorin. Their patience, support, laughter, and love carry me through my journey. They have also unselfishly encouraged and supported the many adventures in my calling as a teacher. I would also like to dedicate this book to my parents, Dolores and Joseph Prantil. They let me find my own way and gave me the wings to follow my dreams. I wish to thank my mentors, Paul Dawson and Anthony Ingraffea at Cornell University. Their expertise and dedication to teaching computational methods led me to pursue its pedagogy with enthusiasm. I also thank James T. Jenkins of Cornell University for his unyielding pursuit of excellent theoretical modeling and its use in validating *all things numerical*. I thank them for, in their own collective words, reminding me to "have fun, keep learning, and to never forget how I thought, how I learned, and how I felt …when I was a student." I dedicate this book to *my* students who doubt, prod, question, and keep me young. We travel through the forest together. Finally, I am forever grateful to my Creator who blesses me every day with a mysterious mix of skepticism, faith, failure, humility, humor, energy, and imagination. *Ego adhuc cognita.*

From Christopher Papadopoulos: I dedicate this book to my family, particularly my parents Kimon and Mary Lou. They provided me with every opportunity to become educated, and all that they have done for me has been motivated by love. I dedicate this book to my sister, Emily, who has inspired me to high academic achievement through her own success, and to Clare, who has been a vital part of my life journey for which I am greatly blessed. I also dedicate this book to my cousin Jim Papadopoulos, who is the lead author of a reference that is frequently cited in this book. I have always admired Jim's keen mind for mechanics, his dedication to teaching, and his persistence on convincing me of his point of view regarding the need to incorporate FEA practice in my teaching. I thank all of the people who have mentored me in various capacities, particularly my thesis advisor Timothy Healey, undergraduate mentors Sunil Saigal and Omar Ghattas, teachers Dolores Stieper, Susan Spaker, and Richard Piccirilli, and collegial mentors Habib Tabatabai, Yiorgos Papaioannou, and Indira Nair. In their own way, each of them has challenged me, has argued with me, has had patience, and ultimately has supported me in ways that have led to my intellectual and professional growth. To my first friends in Puerto Rico, Marcelo Suárez, Jaquelina Alvarez, Basir Shafiq, Walter Silva, Ramón Vásquez, and Robert Acar, and many other colleagues, including Marcel Castro, Bill Frey, Héctor Huyke, Sara Gavrell, Luis Jiménez, Aidsa Santiago, and Genock Portela, gracias por darme la bienvenida y que continuemos a trabajar juntos. Finally, I dedicate this book to my many students, from Cornell, Milwaukee, and Mayagüez, who bring me great joy and pride. You are the ones for whom I ultimately write this book.

From Paul D. Gessler: I dedicate this book to my grandfather, Donald A. Gessler (1932–2013). He not only taught me *how stuff works*, which ignited my interest in engineering, but also never stopped teaching me about how to live life and help others in need, in other words, *how the really important stuff works*. I would like to thank my parents, Timothy and Shelley, my fiancée Elise, the rest of my family, especially brothers Phillip, Peter, and John, and friends and colleagues, especially Marshall Schaeffer and Alex Zelhofer. None of my work would be as it is without their influence, support, and distractions (no matter how unwelcome these distractions sometimes seemed at the time). I would also like to thank my advisor Professor Margaret M. Mathison and the rest of the Marquette University faculty for allowing me to take on this project in addition to my research and graduate coursework. Finally, as always, *soli Deo gloria*.

Vincent C. Prantil, Christopher Papadopoulos, and Paul D. Gessler
August 2013

CHAPTER 1

Guilty Until Proven Innocent

I repeatedly tell students that it is risky to accept computer calculations without having done some parallel closed-form modeling to benchmark the computer results. Without such benchmarking and validation, how do we know that the computer isn't talking nonsense?

Clive Dym
Principles of Mathematical Modeling

If you only make one predictive simulation, it will likely be wrong.

Loren Lorig
CEO, Itasca International

1.1   GUILTY UNTIL PROVEN INNOCENT

One of the many advantages of the finite element method (FEM) is that it is relatively easy to create a model and use the method to run an analysis. Often, for better or worse, the method has become commonplace enough to be seen as a sophisticated calculator. In addition to enhanced computational speed, this is due to the development and preponderance of graphical user interfaces (GUI) used as pre- and post-processors to nearly all commercial finite element software.

Yet a great hazard of FEM is also that, with the aid of commercial software, it can be *too easy* to create a model and run an analysis. The ease of operation can foster "computational complacency" [Paulino, 2000] in validating numerical results. It often appears that the myth that "the computer must be right" is alive and well. While, indeed, algorithms in commercial codes are well debugged and are unlikely to contain programming errors, the user is ultimately responsible for making appropriate modeling assumptions and interpretations of the output.

Hand in hand with complacency is the "optimism bias," in which people tend to believe that they are correct in more situations than they really are [Conly, 2013]. In the context of FEA, even honest users who intend to validate their work might mislead themselves, thinking that results are correct because they appear to correspond to a simple theory that they might be applying inappropriately (for example, out of its bounds of accuracy), or they might be missing a key theoretical idea altogether. Like a cancer, computational complacency and the optimism bias can spread. They can develop into bad habits that thwart the user's comprehension of some *minimal requisite skill set* on which use of the numerical method depends.

However, before exploring this minimum requisite skill set in detail, the user must first realize that he or she should be *skeptical* toward all results of a numerical simulation until demonstrating a sound reason to accept them. In short, we often tell our students—beginning with the first lesson—that, like it or not, algorithmic simulation results are *guilty until proven innocent*.

1.2   WHAT A MINIMAL REQUISITE SKILL SET LOOKS LIKE

Once the analyst understands the need for providing proper input and validating the interpretation of output, he or she is ready to learn the fundamental skills that will enable him or her to perform responsible numerical simulations. To motivate this, we first provide an analogy with driving an automobile.

We can all agree that while a driver need not understand scientifically the vehicle dynamics or the thermodynamics of the combustion engine, any driver must have a basic sense of how the vehicle and engine operate. For example, braking on ice is less effective than braking on pavement; or maple syrup should not be placed in the fuel tank. Of course it cannot hurt to have some theoretical knowledge, such as to understand that braking distance increases roughly as the square of velocity, or in qualitative terms, "disproportionately." That is why driving instructors concentrate on teaching elements of automobile acceleration, cornering, smooth braking, and field of vision rather than the theory of internal combustion engines. Moreover, the instructor should be seasoned to anticipate and correct errors made by the learner. In the end, the student develops some innate feel for what constitutes "good driving," and learns to distinguish between "good" and "bad" maneuvers based on experience.

Likewise, in the realm of FEA practice, we believe what is called for is the development of a "gut feel" for what constitutes correct behavior and good modeling practice. We contend that the minimal requisite skill set on which good FEA practice is based has two principal components:

1. the ability to apply basic theory of Mechanics of Materials; and

2. the ability to make good modeling decisions, including choice of dimension, element type, mesh discretization, and boundary conditions, based on one's knowledge of MoM and previous experience.

These requirements are based on the intimate relationship between FEA and the theory of elasticity, of which a minimal understanding is constituted by classical Mechanics of Materials. They also appeal to pedagogical theory that states that confronting misconceptions—particularly when they are deeply held—is an effective means to eventually enable the learner to overcome them and replace them with appropriate conceptions. This anticipates our further remarks in the next section regarding how to help students confront their misconceptions directly.

We note that the minimal requisite skill set does not contain an in-depth, rigorous, mathematical treatment of the theory underlying FEM. Such rigor, while necessary to program algorithms or as a prerequisite for graduate studies, is not essential to operate and perform finite element simulations and correctly interpret their results. For practical applications of FEA, what is imperative is the ability to distinguish between good and bad methods for interfacing with the tool.

Note To The Instructor

A treatment of the background necessary to use the finite element method effectively is given by Papadopoulos et al. [2011]. Here we argue that a top-down, theory-first emphasis employed in many curricula may not be as necessary as has been thought. We believe that teaching the underlying mechanics can be enhanced by introducing the finite element method as early as an Introduction to Engineering course in the freshman year. We also feel that hand calculations in Statics and Mechanics of Materials can be reinterpreted and made more appealing by emphasizing them as steps used to validate and benchmark numerical simulations. Finally, in an upper division course in finite element theory, one may undertake a deeper learning of how to perform an informed computational analysis under the tutelage, guidance, and support of a seasoned, experienced practitioner.

1.3   THE TEN MOST COMMON MISTAKES

> Computational models are easily
> misused…unintentionally or intentionally.
>
> ───────────────────────────────
>
> Boris Jeremić
> University of California Davis

In accordance with our proposed minimal requisite skill set, we now present a useful list of commonly committed errors in FEA practice. While the advanced user will likely recognize many of these errors (hopefully through direct experience!), the novice who has little or no FEA experience might not fully appreciate their meaning at this point. Nevertheless, they serve as a good preview of issues that will arise, and as a reference to which the novice may return as he or she gains more experience.

Recently, Chalice Engineering, LLC [2009] compiled an assessment of mistakes most commonly made in performing finite element analysis in industrial practice. After 10 years of collecting anecdotal evidence in both teaching undergraduates and advising capstone design projects, we found this list to be nearly inclusive of the most common errors encountered by undergraduate students in their introductory finite element method course. The list published by Chalice Engineering is reproduced here verbatim.

1. Doing analysis for the sake of it: Not being aware of the end requirements of a finite element analysis—not all benefits of analysis are quantifiable but an analysis specification is important and all practitioners should be aware of it.

2. Lack of verification: Not having adequate verification information to bridge the gap between benchmarking and one's own finite element analysis strategy. Test data sometimes exists but has been forgotten. Consider the cost of tests to verify what the analysis team produces, compared with the potential cost of believing the results when they are wrong.

3. Wrong elements: Using an inefficient finite element type or model, e. g., a 3D model when a 2D model would do, or unreliable linear triangular or tetrahedra elements.

4. Bad post-processing: Not post-processing results correctly (especially stress) or consistently. Not checking unaveraged stresses.

5. Assuming conservatism: Because one particular finite element analysis is known to be conservative, a different analysis of a similar structure under different conditions may not be so.

6. Attempting to predict contact stresses without modeling contact: This might give sensible-looking results, but is seldom meaningful.

7. Not standardising finite element analysis procedures: This has been a frequent cause of repeated or lost work. Any finite element analysis team should have a documented standard modeling procedure for typical analyses encountered within the organisation, and analysts should follow it wherever possible. Non-standard analyses should be derived from the standard procedures where possible.

8. Inadequate archiving: Another frequent cause of lost work. Teams should have a master model store and documented instructions about what and how to archive. Again, this is a quality related issue. For any kind of analysis data, normal backup procedures are not sufficient—attention needs to be paid to what information and file types are to be archived in order to allow projects to be retraced, but without using excessive disk space.

9. Ignoring geometry or boundary condition approximations: Try to understand how inappropriate restraint conditions in static or dynamic analyses can affect results.

10. Ignoring errors associated with the mesh: Sometimes these can cancel out errors associated with mistake 9, which can confuse the user into thinking that the model is more accurate than it is. A convergence test will help.

While it may come as no surprise, novice users commit many, if not all, of these errors. But these errors continue to be committed routinely even by advanced users and engineers in industrial practice. As suggested earlier, we attribute this to a lack of a minimal requisite skill set (or an inability to apply such fluently). This lack of understanding is due, at least in part, to *computational complacency* [Paulino, 2000] and the *optimism bias* [Conly, 2013] cited earlier.

Avoiding such errors is not simply a matter of telling and re-telling the student "how to do it." Most students learn by repeated attempts in the face of incorrect reasoning and results. It is through repeated corrections in the face of practice that we learn, not simply by being presented with how things ought to work. Therefore, before a sense of good modeling practice can truly be learned and internalized, the student must come to appreciate the value of being skeptical about initial numerical simulations, i. e., that they are *guilty until proven innocent*. Students must realize and *care* that their intuition might be incorrect. Then they must actively work to deconstruct their previously incorrect model, and replace it with a model with deeper understanding. Likewise, the good instructor must provide a supportive environment in which students are encouraged to explore problems in which they are likely to make errors, and then coach them to be self-critical, to realize and understand the errors that they have made.

Indeed, as suggested by the attention on student misconceptions in the literature on pedagogy [Hake, 1998, McDermott, 1984, Montfort et al., 2009, Papadopoulos, 2008, Streveler et al., 2008], when students are forced to work out a problem with judicious questioning and investigation where their initial reasoning was incorrect—again, in Ken Bain's words, an *expectation failure* [Bain, 2004]—their learning retention is greater, and their recall and critical thinking skills are enhanced. We take up this point further in the last section of this chapter when we recommend our pedagogical strategy for FEA.

1.4 MAN VS. MACHINE

> It's foolish to swap the amazing machine in your skull for the crude machine on your desk. Sometimes, man beats the machine.
>
> David Brooks
> The New York Times

It is noteworthy that many introductory texts for the study of finite element analysis make use of some form or the other of the necessary procedural steps in applying the method in practice. Then students are provided exercises in applying these procedural steps by means of hand calculations. The procedural steps that a typical finite element analysis should include are as follows:

Ask what the solution should look like: An analyst must have some idea of what to expect in the solution, e. g., a stress concentration, and other characteristics of the solution, such as symmetry.

Choose an appropriate element formulation: One needs to understand, from knowledge of the expected solution, what elemental degrees of freedom and polynomial order of approximation are necessary to accurately model the problem.

Mesh the global domain: With knowledge of the expected solution and the chosen order of *interpolation*—the estimation of the solution at a general location based on the computed solution at the grid points of the mesh (the order of which could be linear, quadratic, etc.), one can wisely select a number and arrangement of elements necessary to adequately capture the response.

Define the strain-displacement and stress-strain relations: It is important to know what formulation your commercial software code has programmed into the analysis module. Classical small strain relations are appropriate for linear, static stress analysis. The user must provide a constitutive law relating stress and strain.

Compile the load-displacement relations: The element matrix equations are either derived in closed form *a priori* or computed via numerical integration within the analysis code.

Assemble the element equations into a global matrix equation: This step is performed algorithmically with knowledge of the element degrees of freedom and nodal connectivity. This global equation relates externally applied conjugate forces and associated nodal point degrees of freedom. It represents a generalized form of nodal point equilibrium.

Apply loads and boundary conditions: Because there are multiple prescriptions of statically equivalent loads and boundary constraints, their precise prescription must be justified based on problem symmetry and proximity to internal regions where accurate stress results are most desired.

Solve for the primary nodal degrees of freedom: Solve the appropriately reduced global matrix equation.

Solve for the derivatives of primary degrees of freedom: This involves calculating generalized reaction forces at nodes and strains and stresses within elements.

Interpret, verify, and validate the results: Based on comparisons with initial expectations, experimental data, analytical benchmark results, or other reputable numerical solutions, have the calculated results converged and are they reasonable?

Again, the novice might not completely understand or appreciate the meaning of each step at this time. However, he or she can still gain some sense and insight into the procedure. In particular, it is very telling that the steps break down succinctly into those performed by the analyst and those performed by the computer. Even the novice will appreciate the complementary roles of the human and the machine from the very outset.

With the advent of high speed computers, it is clear that the machine wins in the battle of raw speed and avoidance of computational error. However, while speed and computational accuracy are necessary, they are not sufficient—and not even most important—for producing good FEA results. The machine cannot provide the intellect, strategy, and judgement of the human mind, all of which are crucial to perform good analysis.

The myth that the "computer is always right" comes, in part, from the truth that yes, most commercial finite element software has been sufficiently debugged, removing most or all internal programming errors. Studies by Jeremić [2009] show that programming errors in commercial codes persist only in a very small percentage of cases. In short, the computer, while working fast, also works nearly flawlessly. It can therefore do the "heavy lifting" required to analyze complex problems that lead to the solution of problems with thousands and even millions of degrees of freedom.

But most errors encountered in finite element analysis are either due to incorrect user input, i. e., *garbage in—garbage out*, or due to lack of prudent judgement regarding dimensional approximations, active degrees of freedom, loading strategy, sensitivity to boundary conditions, or the nature of the correct theoretical solution. That is, they can often be traced to one of two causes: incorrect understanding of finite element modeling, or poor application of strength of materials, and often both to varying degrees.

In most cases, therefore, it is operator error to blame for all of the top ten mistakes [Chalice Engineering, LLC, 2009]. To correct these mistakes, the analyst must look for cause and effect. And as remarked, most often, the code is not the cause, although sometimes the user should investigate if the model programmed in the algorithm is, in fact, the correct model for the application at hand.

Thus, when the task at hand can be described in an efficient and robust algorithmic form, the task should be owned by the machine. In those instances where the task requires judgement and/or compromise, the mind trumps the processor. And this is where the practice of numerical analysis most often goes awry. It perhaps comes as no surprise that the *ten most commonly made mistakes are found only in the procedural steps performed by the analyst and none involve the steps performed algorithmically by the computer.* This glaring reality is the driving force behind our novel approach to learning the finite element method wherein we focus on user behaviors rather than on derivations of algorithms.

1.5   PUTTING IT TOGETHER: TOWARD A NEW FEA PEDAGOGY

We have reviewed common errors and standard procedure, in which we emphasize the need for the analyst to be skeptical and to take responsibility for making good judgements. Recalling our overall pedagogical philosophy based on constructivism and encounter of misconceptions, we now outline our vision of a new FEA pedagogy that prioritizes user behaviors. We draw from our own notes and examples to provide a set of exercises and case studies in which students can encounter common errors and expectation failures in a safe environment, and in which they can iteratively address and correct their misconceptions. We promote three effective approaches to ferreting out these misconceptions:

1. utilizing case studies that present commonly encountered expectation failures in students' understanding of mechanics;

2. identifying specific user input, reasoning, or post-processing decisions that result in the specific misunderstanding of the problem at hand; and

3. validation of results, such as by performing repeated convergence studies to verify numerical simulations, comparison with benchmark solutions, or comparison with experimental results.

We strongly believe that for the novice user, it is prudent to focus on the procedural steps that require interaction, judgement, and interpretation, particularly through repeated experience confronting errors and making corrections. This is in contrast to traditional approaches in which a significant amount of classroom time is spent teaching the underlying mathematical formulation of routines that are ultimately performed without error by the computer, such as rote calculation of element stiffness matrices, assembly of global stiffness matrices, and solution of the principal degrees of freedom.

While we think it is important for students to know that such internal computations are made, deriving such procedures should not be done at the expense of providing repeated experiences in which students encounter and correct the errors and misconceptions that we already know they will make. Rather, we believe this time would be better spent on discussions of, say, how stresses vary within and between neighboring elements, and if the modeler's decision captured this behavior correctly. As misconceptions are overcome, and good procedural habits and intuitions are formed, then the student is all the more pre-disposed to learning and appreciating important aspects of the underlying theory at later stages in their education.

In summary, we boil everything down to four concurrent practices.

1. Introduce students to the finite element method much earlier in their curriculum [Papadopoulos et al., 2011], e. g., in elementary Mechanics of Materials.

2. Focus on applications that illustrate and highlight common pitfalls and ways to circumvent them, e. g., choosing proper element formulations, correctly prescribing boundary conditions, and validating solution results.

3. Keep mathematical derivations to a minimum and focus these primarily in areas directly related to mechanics principles, e. g., equilibrium and approximation by interpolation.

4. Highlight a succinct list of commonly accepted good and bad practices in applications of finite element analysis.

We note in closing that there is a growing body of work on what modelers feel is appropriate skepticism with which preliminary simulation results should be judged in both academic and industrial environments. There are a variety of research findings on the teaching of finite element analysis to undergraduates [du Toit et al., 2007], computational complacency [Paulino, 2000], and the reliability of simulation results [Hatton, 1999] which the reader may wish to further explore.

CHAPTER 2

Let's Get Started

Seek the model which is simple, but not too simple.

Albert Einstein

Essentially, all models are wrong, but some models are useful.

George E.P. Box
Professor Emeritus, University of Wisconsin

Note To The Instructor

Here we detail the kind of knowledge, rooted in Mechanics of Materials, that is important for using FEA effectively. While some finite element theory is important, it should not be considered to be a barrier to the early incorporation of FEA in the curriculum; rather, the requisite knowledge is meant to be built throughout the curriculum as the undergraduate student advances. Mechanics educators and practitioners have absorbed some concepts so well that it is easy to forget that these concepts are relatively new to students. Many technical areas must be learned in order to interpret FEA results, catch modeling errors, and guide design. One essential kind of knowledge is comprised of concepts, simplifying physical assumptions, and critical thinking that takes place throughout the undergraduate engineering curriculum. We do not advocate that students learn less mechanics theory. With the advent of powerful analysis tools, we specifically advocate that students should learn as much if not more—a holistic approach that promotes a qualitative understanding of "what affects something else," an expanded grasp of definitions and core concepts.

In the Preface and Chapter 1, we proposed that the kind of knowledge that is important for using FEA effectively falls into two categories:

1. the ability to apply basic theory of Mechanics of Materials to formulate initial expectations of results and related estimates, and to interpret or benchmark results and

2. the ability to make good modeling decisions, including choice of dimension, element type, mesh, and boundary conditions, based on knowledge of MoM and previous experience.

In this chapter we explore the first of these categories, namely the synergy between Mechanics of Materials and Finite Element Analysis. We begin this chapter with a bird's eye view of some qualitative aspects of MoM that the reader should begin to appreciate, followed by a review of what we regard are the minimum essential elements of MoM theory required to undertake study of FEA. We close with two examples that can be solved by hand calculation as a means to illustrate the finite element method.

Some colleagues are concerned that use of FEA in early courses might supplant a strong understanding of Mechanics of Materials principles because the effort normally done by hand can now be done by "pressing a few buttons." We insist that this is neither our point of view nor a circumstance that is likely to occur under a pedagogy that is committed to ensuring that students form good habits of understanding modeling assumptions and validation procedures. We insist that use of FEA requires even *more* theoretical understanding so that it can be applied with skill. The usual adjuration to "calculate problems first by hand" can then be re-interpreted as "take steps to validate and benchmark your FEA solution."

2.1   QUALITATIVE CONCEPTS OF MECHANICS OF MATERIALS

Here we present a list of qualitative concepts that can be read at once by novices and experts, motivated by ideas presented in [Papadopoulos et al., 2011]. While the expert will recognize many of these ideas from experience, the novice can begin to appreciate the qualitative concepts and ideas that a more seasoned practitioner uses with confidence and fluency. We recommend that students periodically return to this list after doing some of the example problems so that they can develop a better feel for how these ideas appear in practice. The presence of this list at the beginning of the chapter should not be interpreted to mean that the student must master this list all at once on first reading. Rather, practice itself is what will help the student to internalize these ideas and develop the fluency of an expert. This list of qualitative concepts is as follows.

• All structures, no matter how strong, are deformable at least to a small degree. This means that when loads are applied, the material points in the structure move or displace. Many structural elements can be modeled as simple springs as a means to understand the relationship of force to displacement in the structure.

• Studying the exact geometry of a structure and its actual displacements under loading can be very complicated with many resulting equations being nonlinear. In many structures of practical interest, however, the displacements will remain small compared to the overall size of the structure, and simple small displacement approximations can be made that lead to simpler, linear relations. Such linearity renders the ability to superpose basic solutions, or to scale any solution in load or overall size.

• One has the ability to interpret a result in terms of basic ideas or elementary asymptotic solutions. For example, the bending moment transmitted by a cross section; the force and moment equilibrium of loads plus reactions; the maximum bending or twisting strain at an outer fiber; and rigid body degrees of freedom of a body or system.

- Stress is a tensor, a directional specification of tractions across arbitrarily oriented surfaces. Principal directions exist on surfaces where the shear stress vanishes. There are no tractions on a free surface, so principal directions are parallel to the surface, and sometimes predictable from symmetry.

- For isotropic material failure, we can ignore stress orientation and use a scalar invariant as a failure metric.

- The source of stress concentrations is based on specific geometric features, such as re-entrant corners or cavities.

- Structures with a single load path are *determinate*, and the resultants are known from the load. Structures with multiple load paths are *indeterminate*, e. g., springs in parallel share the load. Adding material generally increases the load carried by a support, and perhaps even its peak stress.

- Indeterminate structures are often called *redundant*. They obey the laws of static equilibrium, but these equations alone are insufficient to determine the force distribution in the system. Additional equations enforcing compatibility are necessary. These describe how the displacements of material points in the structure must behave in order for the structure to remain intact.

- An idealized pinned support neglects modest moments that exist in the actual physical structure. Similar idealizations hold when modeling other classical localized boundary conditions such as built-in or compliant constraints.

- Analysts must be aware that the world is not rigid, and particularly that prevention of lateral strain is not always realistic.

- When calculating stress, users should exploit St. Venant's principle, i. e., it may be possible to ignore the actual compliance of an end support sufficiently far away from the point of load application.

2.2   THE STRESS TENSOR

In Mechanics of Materials, one is introduced to the basics of stress and strain and their relation in Hooke's Law. Recall that external loads on a structure produce internal forces and moments that result in internal stresses. The concept of stress describes how reactions of the structure to external loads are distributed across arbitrarily-oriented planes in the structure. Recall that there are fundamentally two basic types of stress: (i) normal stress, σ, and (ii) shear stress, τ, as illustrated in Fig. 2.1.

Although it is common to refer to "bending stress," "torsional stress," "bearing stress," "single shear," "double shear," "punching shear," etc., we emphasize that these names do not represent

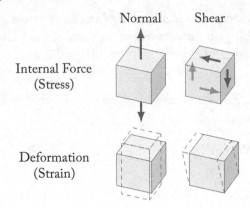

Figure 2.1: The concept of normal and shear stress and strain components is illustrated on infinitesimal volumes.

other basic kinds of stress; rather, they are names assigned to internal stresses specific to commonly studied load cases. All types of stress ultimately can be classified as either *normal* or *shear*. Normal stresses result from:

1. axial loads and deformation of prismatic rods or bars,

2. transverse loads, moments, and the associated curvature in prismatic beams, and

3. approximations of bearing stress.

Shear stresses result from:

1. transverse shear forces and the associated lateral deformations in prismatic beams,

2. torsional loading and rotational deformation in prismatic shafts, and

3. approximations in single shear, double shear, and punching shear.

2.3 IDEALIZED STRUCTURAL RESPONSES

> Theories are like maps; the test of a map lies not in arbitrarily checking random points, but in whether people find it useful to (use it to) get somewhere.
>
> Ronald Giere

Perhaps you have noticed that many of the problems studied in an elementary Mechanics of Materials course consist of highly regular structural forms: rods or bars with uniform cross section; circular shafts and pipes; and beams with uniform and prismatic cross sections. Perhaps you never thought much about just how simple these forms are, but they possess twin properties almost akin to a lucky accident of nature:

1. they possess simple closed form stress-strain and load-displacement relationships that are amenable to hand calculations and

2. they are widely useful and applicable in countless examples of engineering design and construction.

Indeed, the determination of internal stresses in these basic elements follows very simple analysis that is highly accurate. Whether it is obvious or amazing that these common forms should succumb to such simple analysis can be debated by the philosophically inclined. Regardless, this wonderful situation enables engineers to prescribe the use of these objects widely with a high degree of confidence in understanding their behavior. We now review these basic forms in detail.

2.3.1 AXIAL RESPONSE

A long slender bar, subjected only to axial end forces, and whose weight is neglected is a 'two force member' whose internal forces are parallel to the bar itself. Bars are further assumed to undergo small displacements and exhibit negligible out-of-plane effects, i. e., we assume no change in the cross-sectional dimensions as the material element deforms under normal stress. The internal normal stress can be produced by tensile and compressive axial forces, P, that act purely normal to the cross section as shown in Fig. 2.2. The value of this stress, denoted by σ_{axial}, is a normal stress given by the well-known relationship

$$\sigma_{axial} = \frac{P}{A}.$$

In addition, the axial displacement of a long bar of length L under uniform load P is given by

$$\delta_{axial} = \frac{PL}{AE}.$$

2.3.2 LATERAL SHEAR RESPONSE

One way that a shear stress can be produced is by distributing a lateral (or transverse) force, V, in the plane of a cross section, as shown in Fig. 2.3. This stress will not, in general, be uniform over the cross section. However, for certain regular shapes its intensity can be estimated using the well-known formula

$$\tau_{lateral} = \frac{VQ}{It},$$

where V is the resultant of the lateral force vectors, I is the second area moment of the cross section, and Q and t are, respectively, the first area moment of the cross section and thickness

Normal Stress

Figure 2.2: Average normal stress distributions in a bar due to axial load on faces perpendicular to the load.

(or width) of the cross section at the location where the stress is being evaluated. Because the calculation of Q is sometimes involved, an approximation for the maximum shear stress in the section due to this type of loading can be easily obtained by knowing the shape of the cross section, where, for instance

$$\tau_{\text{lateral, max}} = \begin{cases} \frac{4V}{3A} & \text{for circular cross sections,} \\ \frac{3V}{2A} & \text{for rectangular cross sections.} \end{cases}$$

Shear Stress

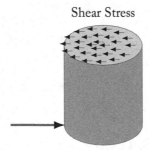

Figure 2.3: Shear traction is distributed perpendicular to the normal of the cross section.

2.3.3 BENDING RESPONSE

Both tensile and compressive normal stresses can also be caused by bending moments, as shown in Fig. 2.4. If the beam has a prismatic section and is symmetric about the transverse plane, the pure bending assumption that 'plane sections remain plane and normal to the neutral axis' can be applied to yield the well-known formula to predict the bending stress at a given distance from the neutral axis:

$$\sigma_{bending} = \frac{My}{I},$$

where M refers to the resultant moment, I represents the cross-sectional property known as the area moment of inertia about an axis passing through the centroid of the cross section, and y represents the distance from the neutral axis toward the outer edge of the cross section where the stress is being evaluated. The displacement of a beam due to a transverse loading can be determined

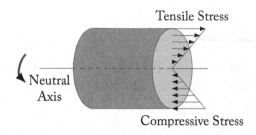

Tensile Stress

Neutral Axis

Compressive Stress

Figure 2.4: Stress distribution due to bending loads varies linearly through the cross section.

by integrating the fourth-order differential equation

$$\frac{d^4 v}{d x^4} = -\frac{w}{EI},$$

where E is the modulus of elasticity and w is the load per unit length applied transversely to the beam. This basic theory of beam bending is often referred to as *Euler-Bernoulli beam theory*.

2.3.4 TORSIONAL RESPONSE

Shear stresses can also develop when a torque is applied to a shaft. If the shaft is circular or annular in cross section, it can be assumed that cross sections remain parallel and circular. From this assumption, the shear stress due to torsion can be predicted at a point at a given radial distance, ρ, away from the center by the well-known formula

$$\tau_{torsion} = \frac{T\rho}{J},$$

where T is the total torque carried by the section and J is the polar moment of inertia of the cross section. These stress components are illustrated in Fig. 2.5. Under these conditions, the axial twist (sometimes referred to as angular displacement) along such a shaft of length L can be calculated from the formula

$$\theta = \frac{TL}{GJ},$$

where G is the modulus of rigidity.

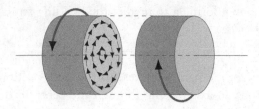

Figure 2.5: Internal stresses due to torsion loads are distributed as shear tractions.

Example 2.1: Simple Truss Analysis

A weight is suspended by three bars as shown in Fig. 2.6. All three bars are made of steel, $a = 16\,\text{in}$, $b = 12\,\text{in}$, $c = 12\,\text{in}$, the diameter of each bar is $0.5\,\text{in}$, and $W = 5000\,\text{lbf}$. Determine the force carried by each bar.

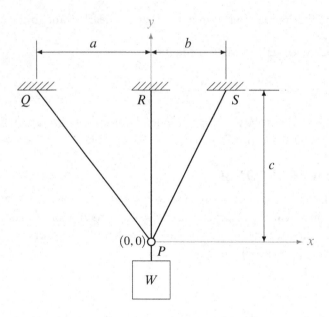

Figure 2.6: A three-bar structure supporting a weight forms an indeterminate truss.

A Free Body Diagram (FBD) of point P reveals that there are three unknown forces, as shown in Fig. 2.7.

Example 2.1: Simple Truss Analysis (continued)

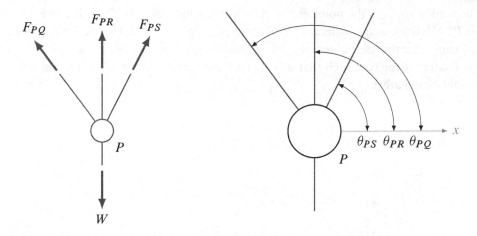

Figure 2.7: Free body diagram of point P with bar angle conventions.

However, there are only two equations of static equilibrium:

$$\sum F_x : \qquad F_{PQ} \cos\theta_{PQ} + F_{PR} \cos\theta_{PR} + F_{PS} \cos\theta_{PS} = 0,$$
$$\sum F_y : \qquad F_{PQ} \sin\theta_{PQ} + F_{PR} \sin\theta_{PR} + F_{PS} \sin\theta_{PS} = W,$$

where the angle for each bar is measured in the counterclockwise direction from the positive x-axis. Such a system is called *statically indeterminate* because the equations of static equilibrium are insufficient to determine the forces in the structural elements. Analysis of a statically indeterminate system requires additional equations that account for the structural deformation, i. e., how the bars deform under their applied load.

Inverting the force-displacement relation from Section 2.3.1, $F = (EA/L)\delta$, where E is the modulus of elasticity, A is the cross-sectional area of the bar, and L is the (initial) length of the bar, allows us to interpret each bar as a spring with equivalent stiffness

$$k = \frac{EA}{L}.$$

Denoting the stiffness of each bar by k_{PQ}, k_{PR}, and k_{PS}, and the deformation of each bar by δ_{PQ}, δ_{PR}, and δ_{PS}, we can rewrite the equilibrium equations as follows:

$$\sum F_x : \qquad k_{PQ}\delta_{PQ} \cos\theta_{PQ} + k_{PR}\delta_{PR} \cos\theta_{PR} + k_{PS}\delta_{PS} \cos\theta_{PS} = 0,$$
$$\sum F_y : \qquad k_{PQ}\delta_{PQ} \sin\theta_{PQ} + k_{PR}\delta_{PR} \sin\theta_{PR} + k_{PS}\delta_{PS} \sin\theta_{PS} = W.$$

Example 2.1: Simple Truss Analysis (continued)

After the load is applied, the point P, which is initially located at $(0, 0)$, will move to a new location P'. We use u and v to denote, respectively, the horizontal and vertical components of the displacement from point P to point P', as illustrated in Fig. 2.8. Note that by convention, we have illustrated the case such that $u > 0$ and $v > 0$, but in general, one or both of these values could be negative.

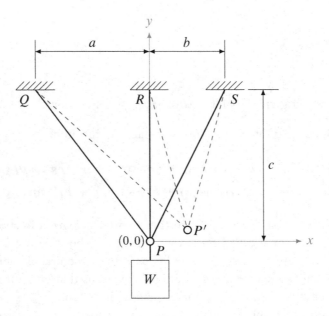

Figure 2.8: The structure deforms and point P displaces as the load is applied.

As suggested by Fig. 2.8, both the length and direction of each bar change after the load is applied. However, under many common circumstances, the displacements are small enough such that the change in direction is negligible. *Therefore we will assume, as an approximation, that each deformed bar is parallel to its original position.* This is illustrated in Fig. 2.9 which shows initial and deformed positions of the bar PS near point P, and how the deformation δ_{PS} is geometrically related to the displacements u and v.

Example 2.1: Simple Truss Analysis (continued)

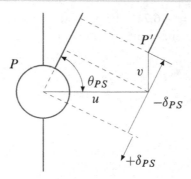

Figure 2.9: The displacement δ_{PS} is comprised of components along x and y directions.

Using basic trigonometry, the deformation of the bar δ_{PS} is related to the displacement of P', (u, v), by the equation

$$-\delta_{PS} = u \cos \theta_{PS} + v \sin \theta_{PS}.$$

Note the negative sign in front of δ_{PS} accounts for the convention that positive δ corresponds to the bar getting longer, but in Fig. 2.8, the bar is contracted.

Because the kinematic description of each bar is standardized (Fig. 2.7), the equations for the other two bars are similar without requiring separate derivations:

$$-\delta_{PR} = u \cos \theta_{PR} + v \sin \theta_{PR}$$
$$-\delta_{PQ} = u \cos \theta_{PQ} + v \sin \theta_{PQ} .$$

These equations are called compatibility equations because the deformations must be *compatible* so that all bars remain connected at point P'. In summary, we now have five equations for the five unknown variables δ_{PQ}, δ_{PR}, δ_{PS}, u, and v. Notice also that these equations are *linear* in these variables. This is a consequence of our use of the approximation that the direction of each bar remains unchanged. For this example, we have in mind that the reader will solve the five equations using a numerical solver such as MATLAB or Excel, and then develop a model of this problem using a commercial FE solver. We recommend assembling the structure using beam or bar elements. Depending on the reader's experience with FEA, it may or may not be clear that both equilibrium and compatibility conditions are simultaneously enforced as part of a *displacement-based finite element analysis*. In our model, using one-dimensional bar (or truss) elements in ANSYS, the finite element method obtains the theoretical solution exactly (up to machine precision): the bar forces are 1287.5 lb in bar PQ, 3197.5 lb in bar PR, and 1456.6 lb in bar PS; the displacements of the loaded point P'

Example 2.1: Simple Truss Analysis (continued)

are $u = 6.00 \times 10^{-4}$ in and $v = -6.74 \times 10^{-3}$ in. The deformed shape can be illustrated by post-processing the finite element results, as shown in Fig. 2.10.

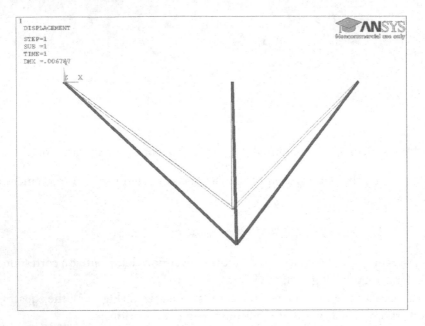

Figure 2.10: The structure deforms and point P displaces as the load is applied. The finite element result matches the exact result for nodal displacements and bar forces.

This example is adapted from Papadopoulos et al. [2013] with permission.

2.4   WHAT DIMENSION ARE YOU IN?

The distribution of stress, strain, and displacement in an elastic body subject to prescribed forces requires consideration of a number of fundamental conditions relating material constitutive laws, material properties, geometry, and surface forces.

1. The equations of equilibrium must be satisfied throughout the body.

2. A constitutive law relating stress and strain must apply to the material, e. g., linear elastic Hooke's law.

3. Compatibility must hold, i. e., the components of strain must be compatible with one another or the strain must be consistent with the preservation of body continuity. This is a critical matter for FEA that is not always discussed in mechanics of materials.

4. The stress, strain, and deformation must be such as to conform to the conditions of loading imposed at the boundaries.

Realistically, all problems are three-dimensional, but satisfying all the conditions outlined above can quickly become intractable. Indeed, closed form solutions to three-dimensional boundary value problems in linear elasticity can be very involved or even impossible. When possible, it is wise to take advantage of simplifications in which the displacement , stress, or strain fields take on a one- or two-dimensional nature. These opportunities afford themselves when a lower dimensional model captures *enough* of the essential behavior.

For instance, in Example 2.1, we tacitly recommended that the 3-bar structure be modeled with beam or bar elements. This was natural enough, but to elaborate, we assumed that behaviors such as lateral contraction of the bars via the Poisson effect, bending, or other stresses not directed along the axes of the bars were negligible. Thus, a model that resembles the behavior of a simple axial bar, and its correspondingly simple behavior as described in Section 2.3.1, is sufficient. It is unnecessary to develop a 'true' three-dimensional model that is more complicated.

In general, when modeling, the metaphor to not 'throw the baby out with the bathwater' is apt. The 'baby' is that which is essential, i. e., the *dominant* mechanics that we choose to keep in the model, such as the dominant axial behavior of the bars in Example 2.1. The 'bathwater' is all of the other mechanics that we choose to neglect, such as the lateral effects in the bars of Example 2.1.

There are several other important situations in which it is appropriate to simplify the dimensionality of a problem. This is evidenced when we realize that simple beam deflection solutions resolve only the deformed shape of the neutral axis of the beam cross section. Indeed, in the simplest beam bending theory that was reviewed in Section 2.3.2, referred to as Euler-Bernoulli theory, the formulae for axial bending stress and maximum deflection are sufficient in the limit as the beam length dominates over the remaining two cross-sectional dimensions. In other words, Euler-Bernoulli beam theory holds only in the limit as the beam becomes "long and slender." The simplest bending relations become progressively more insufficient as the cross-sectional dimensions grow and are no longer *small* compared with the beam's length. In this limit, one can argue that the beam becomes 'hopelessly three-dimensional.'

Other opportunities afford themselves when two dimensions, say in a plane, are either commonly large or small compared with an out-of-plane dimension. In this limit, we have been taught two-dimensional planar solutions for plane stress, plane strain, and axisymmetric conditions. We explore these situations in the following sections.

2.4.1 THE LIMIT OF THE THIN (PLANE STRESS AND PRESSURE VESSELS)

There are many problems of practical importance in which the stress conditions are ones of plane stress. This occurs often in thin members, as shown in Fig. 2.11. In this limit:

1. The stress components σ_x, σ_y, and σ_z do not vary through the thickness, i.e., they are functions of x and y only.

2. Externally applied forces are functions of x and y only.

3. The out-of-plane stress components are identically zero, i.e.,

$$\sigma_z = 0$$
$$\tau_{xz} = \tau_{zx} = 0$$
$$\tau_{yz} = \tau_{zy} = 0.$$

For such cases in FEA, a two-dimensional solid or continuum plane stress element is used.

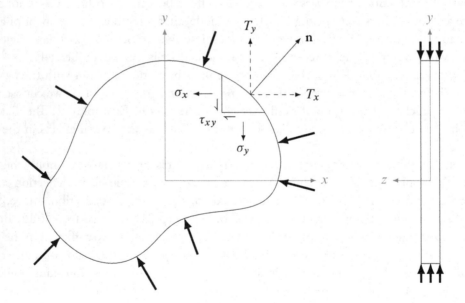

Figure 2.11: A state of plane stress will often result in thin sections with loads applied in the plane.

2.4.2 THE LIMIT OF THE THICK (PLANE STRAIN)

There are many problems of practical importance in which the strain conditions are ones of plane strain. For long, prismatic members subject to lateral loading in the x-y plane, as shown in Fig. 2.12, a state of plane strain will result. In this limit:

1. The strain components do not vary through the thickness, i. e., they are functions of x and y only.

2. Externally applied forces are functions of x and y only.

3. The out-of-plane strain components are identically zero, i. e.,

$$\epsilon_z = 0$$
$$\gamma_{xz} = \gamma_{zx} = 0$$
$$\gamma_{yz} = \gamma_{zy} = 0.$$

For such cases in FEA, a two-dimensional solid or continuum plane strain element is used.

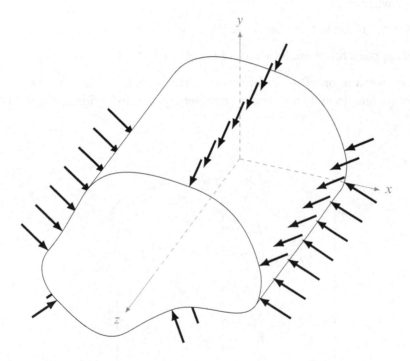

Figure 2.12: A state of plane strain will often result in thick sections with loads applied in the plane.

2.4.3 ANALOGY OF PLANE STRESS AND PLANE STRAIN

For similar cross sections, a solution derived for plane stress is strictly analogous to those for plane strain when using the conversions listed in Table 2.1.

Table 2.1: Conversions of two-dimensional assumptions

Solution	To convert to	E is replaced by	v is replaced by
Plane stress	Plane strain	$E/(1-v^2)$	$v/(1-v)$
Plane strain	Plane stress	$E(1+2v)/(1+v)^2$	$v/(1+v)$

2.4.4   THE LIMIT OF THE ROUND (AXISYMMETRY)

Finally, many practical problems exhibit azimuthal symmetry about an axis. When there is no dependence of the deformation on the angle, θ, in Fig. 2.13, the state of stress will not vary in this direction and the stress and deformation fields reduce to functions of (r, z) only. Such conditions arise whenever:

1. all cross sections in the r-z-plane experience identical deformations;

2. externally applied forces are functions of r and z only; and

3. there is no θ-variation of the deformation in the body, i.e., points in the transverse (r, z) plane always remain in their respective transverse planes following application of the loads.

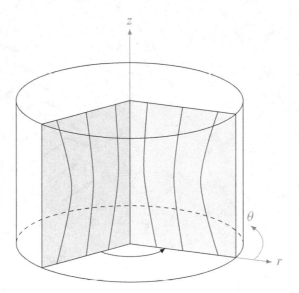

Figure 2.13: An axisymmetric geometry results when there is no variation in the azimuthal (θ) direction.

For such cases in FEA, the body is meshed in the r-z plane and an axisymmetric, two-dimensional continuum element is chosen for the analysis.

2.5   ST. VENANT'S PRINCIPLE

St. Venant's principle, attributed to Barré de St. Venant, is a statement about the change in stress distribution with respect to distance from a prescribed load or boundary condition. St. Venant's principle has significant implications for finite element analysis. It may be stated in a number of equivalent ways.

1. The difference in stresses produced by two sets of statically equivalent forces acting on a surface, A, diminishes with distance from A and becomes negligible at distances large relative to the linear dimensions of A.

2. The detailed distribution of applied forces and moments on a boundary affects the internal stress distribution in the vicinity of those applied forces and moments, but at several characteristic dimensions away from the reactions, the internal stresses are essentially dependent only on the applied external forces and moments, and not on how these forces and moments are applied. A *characteristic dimension* is not an absolute dimension, e. g., "2 in," but rather, is a dimension that is meaningful in proportion to the given system, e. g., "1/3 the width of the bar." This is illustrated in Fig. 2.14.

3. Only stresses in the vicinity of loads are sensitive to the details of how those loads are applied.

4. If self-equilibrating forces act on a surface area, A, the internal stresses diminish with distance from A. The rate at which the stresses attenuate with distance may be influenced by the shape of the body and must be estimated independently in each case.

5. Statically equivalent systems of forces and moments produce the same stresses and strains within a body except in the immediate region where the loads are applied.

6. The localized effects caused by any load acting on the body tend to disappear in regions that are sufficiently far away from the application of the load.

Many of the mathematical representations of the simplest loading conditions are themselves simple. But illustration of the concepts behind these relatively simple formulae are too often lost on students exposed to them for the first time. A powerful teaching tool is the use of quality graphical representations and illustrative examples, both of which appear in Steif [2012] and Philpot [2010]. The reader may also find interesting two handbooks whose focus is a collection of formulae. These references are useful for benchmarking solutions and providing bounding cases used in preliminary analysis [Allain, 2011, Pope, 1997].

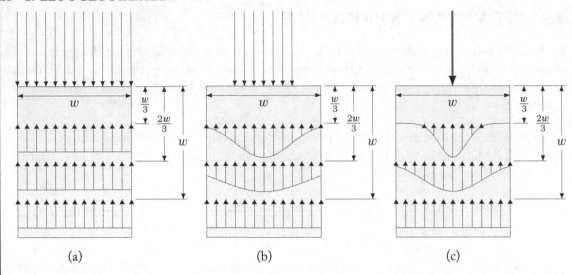

(a) (b) (c)

Figure 2.14: Statically-equivalent sets of applied loads distributed differently over a boundary or part thereof do not alter the internal stresses and their distribution several *characteristic dimensions* (here, measured in terms of the width, w) away from the applied loads. Here, a compression specimen is subjected to equivalent loads (P) over different portions of its ends: (a) full end, (b) half end, and (c) point load. Approximately one specimen width into the bar, the state of stress is a uniform constant stress corresponding to P/A.

2.6 COMBINED LOADING

Note To The Instructor

While students may recognize these idealized loading cases and their respective simple formulae, we often observe that how to linearly superpose these stress components under conditions of even simple combined loading still eludes students even after exposure to the finite element method. Here we consider a simple illustration for which finite element analysis is both straightforward and useful in framing students' hand calculations as benchmarks for simulation results.

SimCafe Tutorial 1: Combined Loading in an Idealized Signpost

The purpose of this case study is to illustrate how combined loading is handled in a straightforward manner using the finite element method. It presents a case study wherein students can perform parametric studies varying the degrees to which the combined loadings are dominated by either axial, bending, torsional, or transverse shear response. It also showcases how internal stresses from combined loads are superposed in a linear analysis.

SimCafe Tutorial 1: Combined Loading in an Idealized Signpost (continued)

Follow the directions at `https://confluence.cornell.edu/display/` `SIMULATION/Signpost` to complete the tutorial.

Example 2.2: Combined Loading in an Idealized Signpost

The cantilevered signpost shown in Fig. 2.15 has dimensions $x_1 = 6\,\text{ft}$, $z_1 = 4\,\text{ft}$, $b_2 = 13\,\text{ft}$, $h_1 = 28\,\text{ft}$, and $h_2 = 8\,\text{ft}$. The system is subjected to the external loads $w_z = 900\,\text{lbf/ft}$, $w_0 = 700\,\text{lbf/ft}$, $F_y = 8000\,\text{lbf}$, and $F_z = $ net weight of the signpost. The signpost is made of steel, and it is assumed that the signpost will remain in its *elastic range*. This means that when the external load is removed, the material will return to its original shape without suffering permanent deformation.

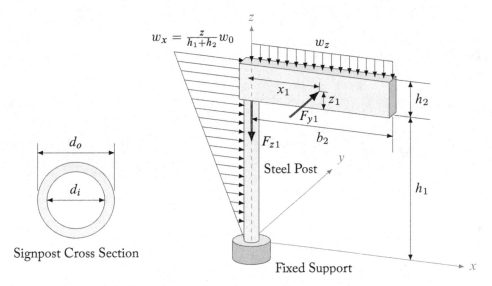

Figure 2.15: Geometrical description of the signpost illustrating dimensions and loads.

The post diameters d_o and d_i must be designed so that the total combined normal stresses and combined shear stresses do not exceed allowable values. Assume allowable stresses of 25 ksi and 16 ksi for normal and shear stress, respectively, which already account for an appropriate factor of safety. This example is adapted from Papadopoulos et al. [2013] with permission and with credit due to Genock Portela.

2.7   A CLOSING REMARK AND LOOK AHEAD

In this chapter we reviewed common structural elements and their usual analyses from Mechanics of Materials. We then used these forms to illustrate broader qualitative concepts and to introduce the finite element method. So far, no major surprises have surfaced, and all of the results are as expected.

As we look ahead to the next chapter, we are now ready to examine problems that have greater geometrical complexity and irregularities. While some aspects of the FEA procedure will be the same as those introduced here, they must now be used with more caution, skepticism, and refinement. Moreover, the user will need to learn some new techniques to completely capture essential details in these new situations.

CHAPTER 3

Where We Begin to Go Wrong

> When all you have is a hammer, everything looks like a nail.
>
> Anonymous

Note To The Instructor

We have often told our students that one of the advantages of finite element analysis is that it is relatively easy to perform. We also add that one of the disadvantages of finite element analysis is that it can be too easy to perform. As ease of use becomes more prevalent, it can belie the complexity of the actual solution to one's problem. Clearly, a distinct advantage has always been to give drudgery and repetitive tasks to the machine to free up time for the analyst to spend critically thinking. So the computer is a fast, but not necessarily intelligent, aid in obtaining sufficiently accurate solutions. The requisite intelligence lives primarily in two places:

1. commercial software's pre-programmed algorithms that approximately model theories with which students may or may not be familiar and

2. an analyst's pre-processing of a model formulation and interfacing this model with the commercial software.

Where students go wrong can often be traced to one of these two lapses in intelligence. The first appears when students attempt problems whose solutions they do not know *a priori*. In such cases, the theory they know may or may not be relevant or sufficient to model the problem. Students often view this as *carte blanche* for initiating a finite element analysis. One common pitfall is that it is more difficult to validate a solution you do not know or understand *a priori*. In such cases, new learners often turn to the theory they know when attempting to validate simulation results. Comparing the results of correct finite element analysis with expectations using inadequate theory is a common mistake made by students in introductory courses. This is particularly true in courses where commercial software is used as part of the student laboratory experience. We illustrate this first "way to go wrong" with three illustrative examples.

3.1   EXCEPTIONS TO THE RULE

If you're running a fever, you will remain home and nurse it…to a point. If your fever reaches 104 °F, however, you may consider visiting your doctor or local emergency room. Analogously, the simple formulae discussed in Chapter 2 can suffer a similar fate in their predictability as one deviates further from the simplifying physical assumptions on which they are based. Take the model for normal bending stress developed in slender beams:

$$\sigma_{\text{bending}} = \frac{My}{I}.$$

This formula is sufficiently accurate when beams are "long and slender;" that is, they are beams in which the length along the neutral axis is large compared with the dimensions of the beam cross section. The transverse deflections under load must also, typically, be orders of magnitude lower than the beam span. As with all good theory, Euler-Bernoulli beam theory is considered valid in a field of geometric dimensions and deformation scales that are bounded by dimensionless ratios. For example, simple beam theory is considered to be applicable when

$$\frac{v}{L} \ll 1; \quad \frac{D}{L} \ll 1,$$

where D is a characteristic linear dimension of the cross section. We may even estimate a range of validity by specifying "lines in the sand" beyond which we can apply the results of the simplified theory:

$$\frac{L}{v} \geq 1000; \quad \frac{L}{D} \geq 20.$$

What is important is that these dimensionless "limits of applicability" are somewhat arbitrary. They serve only as user-defined risk limits in applying simplifying assumptions. They serve as warning posts beyond which we may wish to consider whether the true internal bending stresses are sufficiently modeled by such simple formulae. Of course, the deviations from the simple limit occur gradually as one passes through their range of applicability. Much like the metaphor of a fever, the severity of the dysfunction grows degree by degree. Only finally at "some limit" (that generally varies from person to person) do we decide the formula is *too sick* to be used any further. Like climbers on Mount Everest, if one ignores too many small increments in impending bad weather, one could get caught on the mountain in conditions where equipment suitable for milder weather is no longer appropriate to the task. As the applicability of our simplifications falter, i. e., for sufficiently short beams, the predictions of models based on these simplifications will agree less and less with results observed in practice and in the laboratory.

The moral of the story is simple. The formulae examined in Chapter 2 do not suddenly go bad, no more than a fever jumps from mild to extreme. One tends to step out of the range of applicability of these simple formulae slowly, one degree at a time, until we finally judge predictions based on them to be "sufficiently wrong." Practitioners of FEA must know the applicability of the theories that, in addition to comparison with experimental data, are used to validate any numerical approximation of mechanical behavior.

While the applicability of any simple formula is limited, it is still useful because the range of applicability can generally be large. However, no matter how large the region in which these formulae hold, we must be aware of the fences that bound them lest we utilize poor validation tools to benchmark our numerical simulations.

The following are some of the specific places where mechanics idealizations may either break down or become sufficiently flawed to warrant treading with caution [Papadopoulos, 2008, Papadopoulos et al., 2011]. This list is not inclusive, but we point out several instances where their bearing on validation of FEA is paramount.

- While linearity is applicable for small displacements, it is a poor approximation when displacements grow "sufficiently large." Studying the exact geometry of a structure and its actual displacements under loading can be very complicated with many resulting equations being nonlinear. When this is the case, the advantages accompanying linearity are lost, e. g., the guarantee of unique solutions, ability to superpose basic solutions, and ability to scale any solution in load or overall size.

- Stress concentrations based on geometry such as re-entrant corners or cavities are, in general, not captured by formulae that describe homogeneous states of stress. Stress concentrations are rooted in the interplay of stresses in orthogonal directions and not describable by one-dimensional simplifications.

- For loading that results in fully three-dimensional, inhomogeneous stress states, any and all formulae that rely on lower-dimensional idealizations are often no longer valid.

- When three-dimensional variation occurs, neglect of warpage and lateral strain may not be realistic.

- For loading and geometry that are fully three-dimensional, boundary conditions that are idealized in lower dimensions can no longer be specified in unique terms. There are a variety of approximations to classic boundary conditions such as a clamped support.

3.2   THE LINES IN THE SAND

We do not intend to outline all the boundaries of the simplest theories. This has been undertaken in sufficient detail in many good mechanics of material texts such as Philpot [2010], Steif [2012], and Riley et al. [2007]. We wish here to illustrate a few salient examples. These will serve to highlight what happens in distinct crossings of "lines in the sand," such as:

1. when stress concentrations defy one-dimensional idealization,

2. when previously-insignificant deformation modes become non-negligible, and

3. when geometric dimensions dictate three-dimensional stress states.

3.2.1 A STEPPED AXIAL ROD

SimCafe Tutorial 2: Stress Concentration in a Stepped Axial Shaft

When geometries exhibit discontinuities along a loading path, stress concentrations generally arise. Stress flow is analogous to fluid flow and steep gradients that result in navigating sharp discontinuities result in enhanced stress intensity. What may not be evident is that a discontinuity in geometry requires modeling the geometry in multiple dimensions in order to capture how the stress flows through the domain. Thus, one-dimensional simplifications are not capable of capturing these important effects.

The purpose of this tutorial is to showcase perhaps the simplest stress concentration and point out that it can be resolved in two- or three-dimensions. Simple one-dimensional elements (i. e., simple axial bar elements) that capture constant stress within an element are insufficient to capture stress concentrations, even when many elements are used. In other words, the requisite theory is absent, so mesh refinement is of no utility in converging on the solution. When the element formulation does not contain the necessary physics, h-convergence, or using more elements, captures no more of the solution than does a coarser discretization. This tutorial is meant to highlight where it is relatively straightforward to apply FEA and resolve a solution correctly that belies analytical treatment with uniaxial formulae (such as $\sigma_{axial} = P/A$).

Follow the directions at `https://confluence.cornell.edu/display/` `SIMULATION/Stepped+Shaft` to complete the tutorial.

Example 3.1: A Stepped Axial Rod

Consider a stepped shaft under uniform axial load, P, as shown in Fig. 3.1.

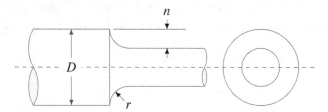

Figure 3.1: Geometrical description of a shaft with a discontinuous step.

Stress concentrations arise due to coupling of the stress response in multiple directions. In the axisymmetric geometry pictured in Fig. 3.1, simplified two-dimensional theory of elasticity can be employed to derive approximate theoretical expressions for the observed

Example 3.1: A Stepped Axial Rod (continued)

stress risers, by fitting such models to experimental data [Solverson, 1953]. Many stress concentration factors fit in this manner are collected in Young and Budynas [2002]. For a stepped shaft with circular fillets:

$$
\begin{aligned}
h &= 3\,\text{in} \\
r &= 1\,\text{in} \\
D &= 8\,\text{in} \\
h/r &= 3 \\
2h/D &= 3/4 = 0.75,
\end{aligned}
$$

a simple fit formula for the axial stress concentration is accurate to within 5% and given by:

$$
\begin{aligned}
K &= C_1 + C_2 \frac{2h}{D} + C_3 \left(\frac{2h}{D}\right)^2 + C_4 \left(\frac{2h}{D}\right)^3 \\
C_1 &= 1.225 + 0.831\sqrt{h/r} - 0.010(h/r) = 2.634 \\
C_2 &= -1.831 - 0.318\sqrt{h/r} - 0.049(h/r) = -2.529 \\
C_3 &= 2.236 - 0.5220\sqrt{h/r} + 0.176(h/r) = 1.8599 \\
C_4 &= -0.63 + 0.009\sqrt{h/r} - 0.117(h/r) = -0.9654 \\
\Rightarrow K &= 1.377,
\end{aligned}
$$

and

$$
\sigma_{\max} = K\sigma_{\text{nom}} = K\frac{P}{A_{\min}} = K\frac{4P}{\pi(D-2h)^2} = 1376\,\text{psi}.
$$

The response of a circular stepped shaft in tension is axisymmetric. An axisymmetric analysis undertaken in ANSYS predicts the stress concentration to within the order of accuracy of the simple formula fit, as shown in Figs. 3.2 and 3.3.

Example 3.1: A Stepped Axial Rod (continued)

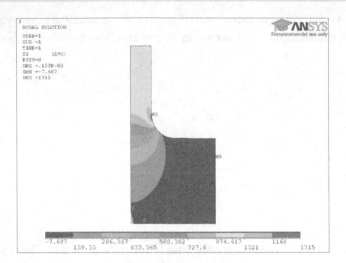

Figure 3.2: The finite element method predicts the axial stress concentration in a stepped shaft.

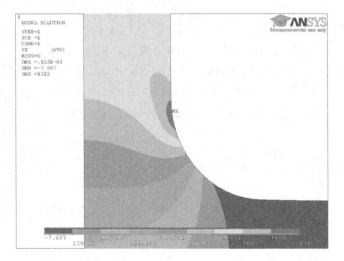

Figure 3.3: The local axial stress concentration is shown in the vicinity of the step fillet.

Because these effects arise from coupling of stress in different directions, one-dimensional theories are incapable of modeling stress concentrations in the vicinity of geometric discontinuities such as re-entrant corners or fillets. Users must be careful to remember that in such cases two- or three-dimensional simulations are required. Because multi-dimensional analysis is required to capture stress concentrations, it is also required in numerical design considerations of how to alleviate such stress risers. For instance, in the case of the stepped shaft, one might ask the question "Is there any way to alleviate the stress concentration at the fillet without changing the diameter on either side or increasing the radius of the fillet?" A three-dimensional analysis reveals that this is actually possible by undercutting the larger diameter portion of the shaft in the vicinity of the original step, as shown in Fig. 3.4.

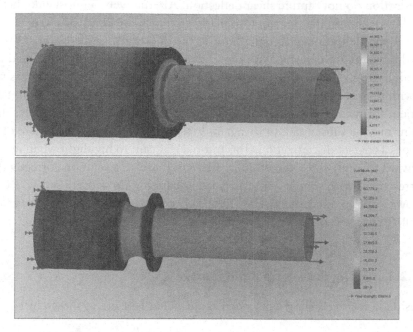

Figure 3.4: It is possible to alleviate a stress riser without changing either diameter of a stepped shaft. A multi-dimensional finite element analysis is required to capture these phenomena. This solution is reproduced from [Papadopoulos et al., 2011] with permission, with particular credit due to Jim Papadopoulos.

3.2.2 A SHORT, STUBBY BEAM

Euler-Bernoulli beam theory, as introduced in strength of materials courses, accounts for transverse deflection due to bending only. Bending deflections can be said to dominate the deformation response when the span-to-depth ratio of the beam exceeds, say, 15. For progressively shorter beams, the assumption that shear deformation can be neglected when compared with the bending deformation is no longer warranted. In these limits, the shear deformation should be taken into account. Timoshenko beam theory accounts for explicit contributions of deformation due to shear. This is the theory one should apply when the span-to-depth ratio of the beam falls below some prescribed limit.

SimCafe Tutorial 3: Stress and Deflection in a Timoshenko Beam

The purpose of this tutorial is to showcase where simple beam theory begins to break down. In some commercial codes, simple one-dimensional cubic beam elements that capture bending deflection do not capture shear deflection. Alternatively, Timoshenko beam theory may be used by default in the element formulation (as with the BEAM188 element in ANSYS v14). When shear deflection is accounted for in the one-dimensional element formulation, results for the beam's tip deflection will not agree with tip deflections predicted by simple Euler-Bernoulli beam theory when the beam is relatively short. Again, attempts to capture this effect with h-convergence will ultimately fail when the necessary physics is not contained in the element formulation. When it is and the results are compared to simpler theory, the disagreement may be substantial. Once again, h-convergence captures no more of the solution than does a coarser discretization. This tutorial is meant to highlight when it is relatively straightforward to apply three-dimensional FEA and resolve a solution correctly that belies analytical treatment with simple formulae (such as bending tip deflection $v = PL^3/3EI$).

Follow the directions at `https://confluence.cornell.edu/display/SIMULATION/Stubby+Beam` to complete the tutorial.

Example 3.2: Large Depth-to-Span Ratio Beams

Consider a relatively short tip-loaded cantilevered I-beam, as shown in Fig. 3.5.

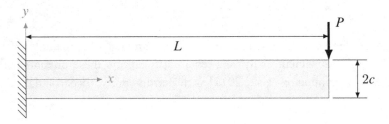

Figure 3.5: A simple cantilever beam is loaded under transverse point tip load P.

The behavior of relatively short beams can be numerically approximated by either one-dimensional beam elements that account for shear deflection or a fully three-dimensional analysis. One should note, however, that while one-dimensional Timoshenko beam elements have interpolation functions for shear deformation, they do not capture the complete three-dimensional state of stress within the beam. For instance, in short cantilever beams the

Example 3.2: Large Depth-to-Span Ratio Beams (continued)

normal stress component at the clamped edge can no longer be predicted with the simple bending formula in Chapter 2.

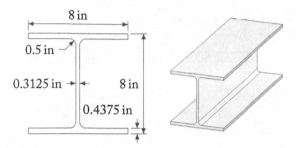

Figure 3.6: Cross section of a short I-beam and a corresponding three-dimensional solid model that can be imported into many commercial finite element software packages.

The solid model is meshed for an I-beam whose span is 24 in. With a span-to-depth ratio of only 3, the actual deformation and stress response will not be modeled well by Euler–Bernoulli beam theory. Three-dimensional finite element simulations indicate that the shear deflections are on the order of those from simple bending theory and the wall normal stresses deviate substantially from those predicted by simple bending theory. Typical contours of displacement and stress for the three-dimensional model are shown in Figs. 3.7 and 3.8 for a tip load of 1000 lb.

Example 3.2: Large Depth-to-Span Ratio Beams (continued)

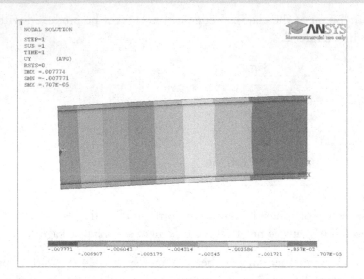

Figure 3.7: Both shear and bending contribute to the total transverse deformation of short beams.

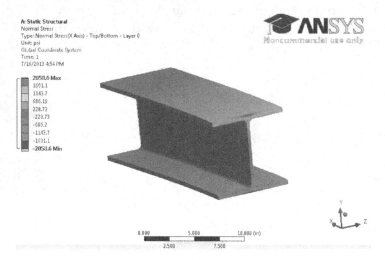

Figure 3.8: The axial stress at the fixed wall deviates substantially from that predicted by one-dimensional beam theory.

Additional three-dimensional models can be run to examine the effects of the length of the beam. Such analyses verify the dependence of both the tip deflection and normal wall stress on the beam's span-to-depth ratio, as evidenced by results in Figs. 3.9 and 3.10.

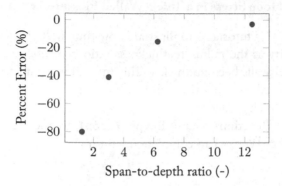

Figure 3.9: The tip deflections in short beams predicted by Euler-Bernoulli beam theory become progressively inaccurate for relatively short beams.

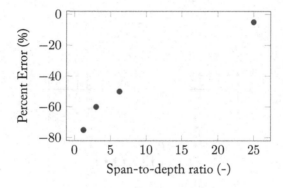

Figure 3.10: The normal stress at the fixed end predicted using Euler-Bernoulli beam theory in short cantilever beams can underestimate the actual normal stress substantially.

3.2.3   A THICK-WALLED PRESSURE VESSEL

The simple formulae outlined in Chapter 2 represent nearly all states of uniform or linearly varying stress. Radial and hoop stresses in pressure vessels become uniform through the thickness as the radius-to-thickness ratio becomes large. Because these formulae are simple and because the variation of both radial and hoop stress becomes nonlinear for thick vessels, analysts may be tempted to push the limits of the simple formulae. Here we point out that, as with the other simple formulae, the deviation from the uniform stress state occurs gradually. When the radius-to-thickness ratio falls below 10, errors arising from predicting stresses with thin-walled formulae become appreciable and thick-walled formulae become increasingly necessary.

SimCafe Tutorial 4: Hoop Stress in a Thick-Walled Pressure Vessel

The purpose of this tutorial is to illustrate how thin-wall pressure vessel theory gradually loses applicability as the radius-to-thickness ratio decreases. As before, this happens gradually as the vessel walls become thicker. This tutorial is meant to highlight where it is relatively straightforward to apply three-dimensional or axisymmetric FEA and resolve a solution correctly for thick-walled vessels.

Follow the directions at `https://confluence.cornell.edu/display/ SIMULATION/Pressure+Vessel` to complete the tutorial.

Example 3.3: A Hydraulic Test Stand

Consider a hydraulic pressure vessel used to apply loads to experimental fixtures in an undergraduate statics and strength of materials laboratory, as shown in Fig. 3.11.

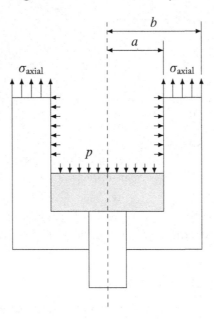

Figure 3.11: Hydraulic test stands are typically moderately thick-walled pressure vessels.

Consider that the pressure vessel is verging on the limits of the thin-wall theory. The outer diameter is 4 in with an inner diameter of 3 in and a 0.5 in wall thickness, giving an

Example 3.3: A Hydraulic Test Stand (continued)

average radius-to-thickness ratio of 3.5. Exploiting symmetry, an axisymmetric analysis of half the vessel is created. The vessel is internally loaded with a constant pressure of 1000 psi. The axisymmetric deformed mesh and internal stresses indicate a stress riser in the bottom of the tank where membrane and bending stresses coincide, as shown in Fig. 3.12.

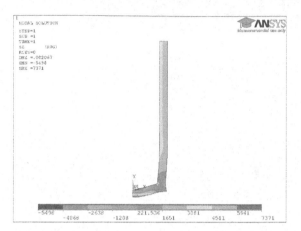

Figure 3.12: Pressure vessel hoop stress maximum occurs in the bottom of a thick-walled vessel.

Far from the discontinuity of the vessel corner, the hoop and radial stress variations in the axial direction in the cylinder wall vanish, as shown in Fig. 3.13.

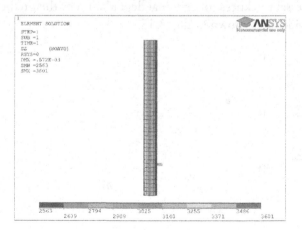

Figure 3.13: Pressure vessel hoop stresses are no longer uniform through the wall of a thick-walled vessel.

Example 3.3: A Hydraulic Test Stand (continued)

Paths through the domain may be defined in many commercial finite element software packages. Here, the variation of hoop stress through the wall thickness is not negligible. The results shown in Fig. 3.14 show the maximum value on the inner diameter predicted correctly by thick-wall theory.

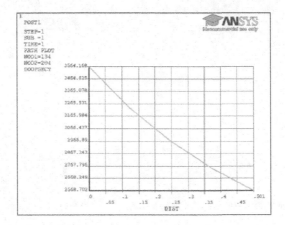

Figure 3.14: Radial variation of hoop stress in the uniform section of the cylinder wall shows a peak value at the inner wall that is underestimated by thin-wall theory.

When we vary the vessel thickness, the gradual degradation of the predictions using thin-walled formulae become evident, as shown in Fig. 3.15.

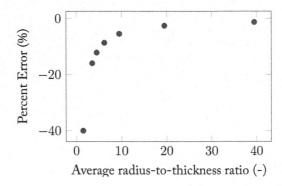

Figure 3.15: Hoop stresses in thick-walled pressure vessels are underestimated by relations based on thin-walled pressure vessel theory.

3.3   UTILITY OF THE FINITE ELEMENT METHOD

The deviations from the simplest stress states that occur in the examples of Sections 3.1 and 3.2 are easily handled by the finite element method. Deviations such as stress concentrations may be predicted using approximate formulae, but these are almost always dependent on details of the specimen geometry; FEA, in contrast, is simple enough to apply in all of these cases and does a good job predicting the correct behavior for elastic deformation and stress. As the finite element method becomes more pervasively used in industry, we feel there is utility in introducing the method earlier in engineering curricula [Papadopoulos et al., 2011]. Distinct advantages to introducing the method throughout one's undergraduate studies include reducing the drudgery and potential errors of computation, focusing on the theory of mechanics, while enabling students to approach more complicated problems that escape the realm of closed-form solutions.

Now recall our earlier point that when using pre-programmed software, the majority of the errors and their severity are attributable to the user. These include faulty input, poor modeling, poor pre-processing, and ignorance of the software protocol. Analogous errors of using a wrong formula or remaining ignorant of a key formula can occur when using hand calculations [Jeremić, 2009, Papadopoulos et al., 2011, Prantil and Howard, 2007, 2008]. The potential for such error in problems like the ones in this chapter is high because the theoretical solutions are likely beyond what most undergraduate mechanical engineering students have learned.

Here FEA can be very beneficial to allow students to explore behavior beyond their basic theoretical knowledge, and it can serve as a bridge for them to discover more advanced theoretical treatments that appear, such as Gieck and Gieck [2006] and Young and Budynas [2002]. Such books are good references for finite element analysts to have at hand for validating numerical solutions for problems whose analytical or empirical solutions have been determined. Using these solutions as benchmarks for FEA analyses helps reinforce the practice of finding published and verified solutions for comparison with numerical simulations. This further underscores our earlier point that we advocate early introduction of FEA in the curriculum, even when it appears to precede the students' current level of engineering knowledge [Papadopoulos et al., 2011].

While applying the finite element method in these cases is relatively straightforward, for more complex geometries and boundary conditions, the prescription of model details leads to situations in which it can become progressively easier for analysts to go wrong applying the method. We discuss illustrative case studies for two such boundary value problems in Chapter 4.

CHAPTER 4

It's Only a Model

A model is a lie devised to help explain the truth.

Anonymous

The truth is always too complex.

Bruce Irons and Nigel Shrive
The Finite Element Primer

Note To The Instructor

The second lapse in intelligence in applying the finite element method occurs when users understand the problem they want to solve, and understand the theory that they believe holds for the problem at hand. The issue is whether the analyst properly poses the finite element formulation of the problem. These types of errors can occur when analysts pre-process a model and

1. apply loads or boundary conditions incorrectly,

2. use an inadequate element formulation for the solution desired, or

3. analyze the problem in an inappropriate dimension, i. e., pose the problem as two-dimensional when three-dimensional analysis is required.

In this scenario, the user falls prey to an old adage wherein the computer is doing what they tell it to do rather than what they want it to do. Here we pose two deceivingly simple problems that cause new learners to often make these common mistakes in problem formulation.

4.1   THE EXPECTATION FAILURE

> We expect regularities everywhere and attempt to find them even where there are none. Events which do not yield to these attempts we are inclined to treat as "background noise," and we stick to our expectations even when they are inadequate.
>
> Karl Popper
> Conjectures and Refutations: The Growth of Scientific Knowledge

As we mentioned in the Preface and elsewhere, we strongly believe in using *expectation failures* as part of our teaching strategy. Because they are crucial to the examples in this chapter, we repeat the words of Ken Bain to remind the reader of their meaning and importance:

> Some of the best teachers want to create an expectation failure, a situation in which existing mental models lead to faulty expectations. They attempt to place students in situations where their mental models will not work. They listen to student conceptions before challenging them. They introduced problems, often case studies of what could go wrong, and engaged the students in grappling with the issues those examples raised [Bain, 2004].

Among the list of common errors made in FEA practice, in this chapter we address misconceptions regarding either

1. the real physics governing the problem or

2. the construction of the finite element model approximating these physical mechanisms.

So an analyst harbors some misconception regarding underlying physical phenomena or details of an appropriate numerical approximation. But, and this is critical, they begin to assure themselves that they do understand. Perhaps they do not remember that "the truth is always too complex" and either our broad simplifications of reality (the simple formulae) or the finite element model approximations (say, lower-order interpolation finite elements) are insufficient for the problem at hand. As we discussed in Chapter 3, analysts will proceed as if these simplifications adequately represent the real behavior. In these cases, analysts may trust the incorrect numerical analysis. Even when presented experimental evidence that does not validate the computational results, analysts can still "cling with fervor" to these incorrect results. Such *computational complacency* may be born of rationalizing that because the software has more theory programmed into it than the user has learned, the computer is more likely right.

In finite element analysis, expectation failures can arise in the following ways.

1. One prescribes boundary conditions that either over- or under-constrain the boundaries by

    (a) not removing all rigid body translation and rotation or

    (b) overly constraining degrees of freedom along a particular direction that preclude deformation and Poisson effects in orthogonal directions.

2. One chooses inappropriate finite element formulations, such as

    (a) planar or one-dimensional elements that are not appropriate for the observed behavior,

    (b) finite elements with inadequate degrees of freedom, or

    (c) finite elements with inadequate order of interpolation.

3. Lower-order interpolations appear to predict behavior more accurately than higher-order interpolations.

4. Meshes with fewer active degrees of freedom appear to predict more accurately than meshes with more active degrees of freedom.

We wish to illustrate these points with two examples where finite element modeling can go wrong. Remember, whether or not simplified theory is appropriate, incorrect finite element results are typically cases of analyst error.

4.2   PHILOSOPHY OF MATHEMATICAL MODELING

> The great masters do not take any model quite so seriously as the rest of us. They know that it is, after all, only a model, possibly replaceable.
>
> C.S. Lewis

> The game I play is imagination in a tight straightjacket. That straightjacket is called the laws of physics.
>
> Richard Feynman

S.L. Hayakawa is noted for pointing out that "the symbol is not the thing symbolized; the word is not the thing; the map is not the territory it stands for" [Dym, 2004], echoing Richard Feynman who recalled that his father "knew the difference between knowing the name of something and knowing something" [Public Broadcasting System–NOVA, 1993]. When engineers attempt to formulate models for systems and processes, it is incumbent upon us to remember that the process, the system *is* "the thing," "the territory." The model is a symbol, word, or map that in some way names the thing. They are not the same. To model some process well requires recasting its real nature into a simplified shell that allows its basic nature to be captured in mathematical form, a set of equations whose solutions tell us something about how the model system behaves under a given set of controlled conditions. An abstraction of the process is shown in Fig. 4.1.

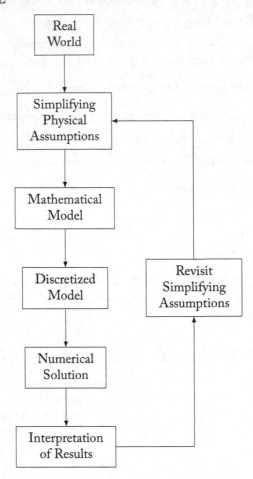

Figure 4.1: A mathematical model is devised by sufficiently simplifying a problem statement such that its formulation can be cast in equation form.

Similar conceptualizations have been illustrated elsewhere and these overviews of modeling are well worth reading: Carson and Cobelli [2000], Dym [2004], Greenbaum and Chartier [2012]. In order to numerically model a system, we must observe the system in nature. We must:

1. collect all information relevant to how the system behaves,

2. detail what we need to find out or predict,

3. specify how well we need to know or predict this behavior, and

4. seriously ask a singularly important question: "What do we expect to happen?"

We should have a knowledgeable, informed expectation of how the system will respond to disturbances, excitation, or loading based on practical experience, prudent observation, and one's understanding of the relevant physics.

In any given process, only a few physical mechanisms tend to dominate the behavior. Making physically simplifying assumptions means deciding what physical mechanisms to retain (recall the baby) and which to neglect (recall the bathwater). The modeler needs to retain the dominant physics and neglect all higher-order effects, making the model as simple as possible, but no simpler. Making appropriate simplifying assumptions is an art whose mastery comes only gradually with continued experience.

After appropriate simplifying assumptions are made, application of a conservation or balance principle results in a differential equation for the boundary value problem. Finite element methods provide a piecewise approximation to the solution of this differential equation. In constructing finite element models, the major inputs from the user are

1. the choice of finite element, which dictates the incremental solution interpolation between nodes and

2. the specific prescription of boundary conditions for the global domain.

We've already learned that beam behavior can be approximated using one-dimensional and three-dimensional models. Here we will use both and compare the results to experimentally measured values.

Recall that boundary value problems are described fully by a governing differential equation coupled with an admissible set of appropriate boundary conditions. For static analyses, the boundary conditions must remove all rigid body translations and rotations.

Upon applying admissible boundary conditions, we solve for displacements throughout the global region. Most commercial finite element software then post-processes the displacement solution to compute

1. reaction forces corresponding to applied displacement constraints and

2. internal stresses which may be displayed or contoured.

One goal in model development is to start with the simplest approximation that captures the physics and provides perhaps crude, but reliable qualitative predictions of system behavior. We will seek to iterate on the model to provide more quantitative results, and then to validate the numerical predictions with experimental observations and test results. All models are approximations whose errors most commonly arise from

1. expectation failures,

2. faulty simplifying assumptions,

3. poor discretization of the domain,

4. poor choice of element interpolation function,

5. incorrect post-processing, or

6. misinterpretation of results.

To validate a numerical solution, it is prudent to perform initial benchmark solutions on representative problems with simplified geometries and boundary conditions. Preferably, these problems are ones whose solutions are known either in closed form or bounded by analytical solutions from above and below. Beyond this, all system modeling employing numerical simulation requires model iterations. Based on previous results, subsequent analyses must be entertained that:

1. relax simplifying assumptions,

2. refine the discretization, or

3. employ higher-order interpolation between solution grid points.

Such model iterations must be performed until the solution converges and independent validation is achieved.

Finite element analysis is a numerical approximation in which the global solution to a large-scale boundary value problem is approximated by a series of finite range functions that are themselves lower-order Taylor series approximations that approximate the local behavior of the solution with sufficient accuracy. These local representations of the solution are based on the Lagrange polynomial interpolation functions that characterize each finite element.

4.3   THE ART OF APPROXIMATION

Modeling and the approximations made therein are an art. When devising numerical approximations on top of the requisite simplifying assumptions, any model is never, strictly speaking, correct, but (hopefully) correct enough. Nearly all numerical approximations in finite element modeling are approximations to theoretical solutions characterized by high levels of continuity and differentiability. But these approximations consist of piecewise, lower-order Lagrange polynomial fits between grid points at which nodal equilibrium is explicitly satisfied. The levels of continuity in displacement sacrificed in the weighted residual are the inherent penalty for the approximation that allows average solutions to continuous differential equations to be obtained from simpler algebraic matrix equations. In some crude sense, numerical analysis is *the fine art of lying by approximation*.

The concept of piecewise polynomial interpolation of a solution over a finite domain is rooted in appropriately truncated Taylor series expansions. In some defined neighborhood of the nodes, a continuous function has an infinite number of truncated Taylor series approximations. The applicable neighborhood over which each series is considered valid then depends on the order of the truncation.

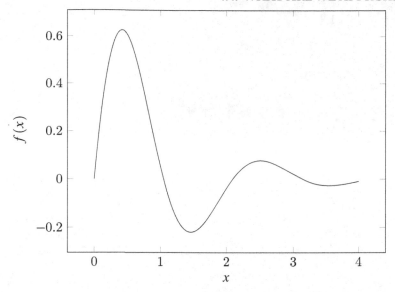

Figure 4.2: A generic function is shown over some global domain.

Consider the function $f(x) = e^{-x} \sin(3x)$, plotted in Fig. 4.2. In the vicinity of the point $x = 2$, the second-order Taylor series approximation represents the function with some level of accuracy in some prescribed neighborhood of $x = 2$, as shown in Fig. 4.3. The linear, first-order Taylor series approximation is a reasonable representation over yet a smaller window. The zeroth-order Taylor series allows for no interpolation. Then it follows that the neighborhood over which an element's interpolation function approximates a known solution with acceptable accuracy will determine the appropriate element size you want in your discretized domain. Therefore, it follows that you cannot know how to best discretize your domain without knowing what element inter-polation, i. e., element type, you have chosen. As we will see, how well higher-order derivatives of these interpolating functions represent the derivatives of the actual solution must be considered in order to determine the accuracy of the stresses predicted by the numerical model.

4.4 WHAT ARE WE APPROXIMATING?

The primary solution variables in FEA are displacements at discrete grid points we call nodes. A discrete solution using the finite element method always delivers an approximate overall solution in the entire domain characterized by

1. maintenance of force equilibrium at all nodes and

2. sacrifice of inter-element force equilibrium in neighboring finite elements that share par-ticular nodal points.

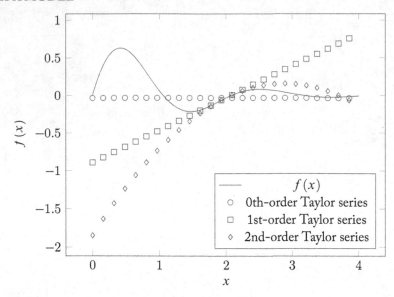

Figure 4.3: Progressively higher-order truncated Taylor series approximations to an arbitrary function model the function's behavior well over progressively larger local neighborhoods.

The displaced configuration of an elastic body is precisely the set of nodal point displacements superposed on the original undeformed configuration. The deformed body acts as an elaborate three-dimensional spring that, upon unloading, would return instantaneously to its original size and shape. The set of nodal point displacements comprise a set of coefficients that each multiply basis functions whose collected weighted sum represents an approximation of the continuous displacement field in three dimensions. Finite element analysis is, in one sense, a piecewise Lagrange polynomial interpolation of this continuous field into many lower-order polynomials whose continuity requirements at nodal points are dictated by the order of truncation of the local Taylor series. It is, therefore, the order of the interpolation or shape function that dictates the variation of displacement along the interior of the finite element.

Now let's consider an idealized finite element analysis as an example of:

1. developing and solving a mathematical model,

2. showcasing where particular errors made in finite element practice might occur, and

3. illustrating where theory embedded in finite element formulations is no guarantee that using finite element analysis will result in an accurate simulation.

SimCafe Tutorial 5: Four-Point Bend Test on a T-Beam

The purpose of this case study is to showcase how the manner in which boundary conditions are applied can change with the number of dimensions in the analysis. Prescription of a single unique "appropriate" set of boundary conditions may no longer exist in a three-dimensional model vs. its one-dimensional analog. In the case study described here, multiple prescriptions of a "simple support" lead to significantly different predicted bending stresses even in the fairly benign circumstances encountered in a four-point bend test.

Follow the directions at `https://confluence.cornell.edu/display/ SIMULATION/T-Beam` to complete the tutorial.

Example 4.1: Four-Point Bend Test on a T-Beam

Consider that we are examining a long, slender T-beam loaded at two symmetric locations on its top surface while being simply supported at its ends along triangular knife-edge supports as shown in Figs. 4.4 and 4.5. The load was applied with a hydraulic cylinder apparatus. Strain gages mounted at several locations between the loading points (where the moment was constant and the transverse shear force was zero) were monitored during the test. We know that the beam is made of isotropic steel with a span of 30 in and constant cross-sectional properties. We wish to accurately predict its peak bending stress.

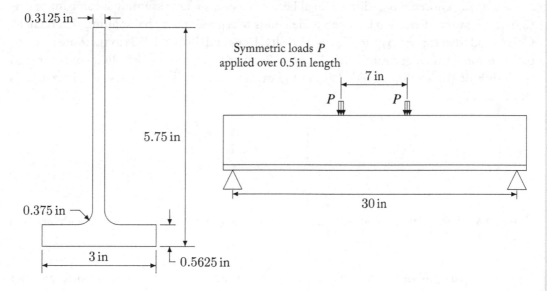

Figure 4.4: A T-section beam cross section is pictured, along with a schematic of the loads applied in a four-point bend test.

Example 4.1: Four-Point Bend Test on a T-Beam (continued)

Figure 4.5: The T-section beam is simply supported along triangular knife edges at each end.

We assume the load is quasi-static. The material remains in the elastic range, the beam is long and slender *enough* for Euler-Bernoulli beam theory to be a sufficient representation of the deformation and internal stress response. We neglect contributions to the deformation from shear deflection. We assume the vertical transverse loads from the hydraulic press can be modeled as pressures over small contact patches. We also assume the simple support at the ends of the beam constrain the transverse displacements at the beam's bottom flange in contact with the knife-edge support.

Having chosen a one-dimensional beam element, we are assuming a cubic interpolation of transverse deflection between node points to represent a global solution that is cubic. One would then expect to generate exact results [Irons and Shrive, 1983] as there are no truncation errors in the approximation. A linear distribution of normal, bending stress through the depth of the section would then be the expected result. The simplest discretization is shown in Fig. 4.6.

Figure 4.6: A one-dimensional finite element mesh using beam elements is loaded with idealized point loads.

Comparisons of the normal bending stress results of the one-dimensional analyses with those determined from strain gage test data from the lab allowed for some interesting comparisons, as shown in Table 4.1. Here we report the stresses in dimensionless form where

Example 4.1: Four-Point Bend Test on a T-Beam (continued)

the actual stress is *normalized* with respect to the *characteristic bending stress*

$$\hat{\sigma} = \frac{PLh}{2I}.$$

This illustrates a further point about the finite element method. It is entirely devoid of any reference to the chosen system of units. These are entirely at the discretion of the user. One need only prescribe a consistent set of units in order to interpret results meaningfully. Because the units are discretionary, results from linear static analyses scale linearly with load and dimensionless results are rendered independent of the actual specific load, section properties, or material constants chosen.

Table 4.1: Results of one-dimensional beam analyses

	$\sigma_{\text{bottom}}/\hat{\sigma}$	$\sigma_{\text{top}}/\hat{\sigma}$
Experiment	0.1108	-0.2464
Euler–Bernoulli beam theory	0.1134	-0.2724
FEA: beam elements	0.1134	-0.2724

Based on these results, in which physical experiment, simple beam theory, and finite element simulation are in good agreement, we could conclude that we have obtained an accurate answer for the peak bending stresses in the beam, and in particular, that the use of beam elements is an appropriate choice for the finite element model. We further comment that the alert reader should surmise that the peak stresses occur at the midpoint ($x = 15$ in).

Recall that models are approximations of reality, and it is quite possible that more than one model is capable of producing an accurate result. It is well worth asking if a fully three-dimensional analysis would *also* verify these results. This may, in fact, be what one *expects* at first glance.

To investigate this question, a three-dimensional model is created in which the hydraulic loads are approximated as pressure loads over the small contact areas. We also assume that the knife-edge supports at the left and right ends can be modeled by constraining the transverse (z-direction) and out-of-plane (y-direction) displacements at all points along left and right edges of the beam's bottom flange (that is, along the edge lines parallel to the y-direction), as shown in Fig. 4.7.

Example 4.1: Four-Point Bend Test on a T-Beam (continued)

All translational DOF fixed
for all nodes at the supports

Figure 4.7: Simple support boundary conditions used in the finite element model are applied throughout the cross section.

Preliminary results indicate that the stress variation is fairly linear through the cross section, as depicted in Fig. 4.8.

Figure 4.8: Axial stress distribution in the three-dimensional beam model varies linearly through the section.

But the normal bending stress at the extreme fibers predicted by the three-dimensional finite element model does not agree with experiment as outlined in Table 4.2. The stresses are under-predicted on top by 11% and on bottom by 52%.

Example 4.1: Four-Point Bend Test on a T-Beam (continued)

Table 4.2: Results of one- and three-dimensional beam analyses

	$\sigma_{\text{bottom}}/\hat{\sigma}$	$\sigma_{\text{top}}/\hat{\sigma}$
Experiment	0.1108	-0.2464
Euler-Bernoulli beam theory	0.1134	-0.2724
FEA: beam elements	0.1134	-0.2724
FEA: solid elements	0.0536	-0.2184

At this point we have arrived at an expectation failure, for the results from our "obviously correct" three-dimensional model do not match the accepted values from the previous analysis. In the spirit of our pedagogical approach, it is now incumbent upon the student to speculate as to why this has occurred, and likewise, it is imperative that instructors supportively coach their students toward a more correct understanding of the solution. For instance, some possible reasons for our discrepancy include the following.

- The simple beam calculations were made with a cross section that neglected the fillets between the web and flange. The solid-element model includes the fillets, resulting in a stiffer structure. The error introduced by neglecting the fillets is less than 0.5% for this geometry.

- Experimental errors, including reading of the applied pressure, locations of the supports and load application points, inaccurate modulus of elasticity, and strain gage errors, caused the measured strains to be inaccurate. If only the three-dimensional model were being compared to the experimental results, this might have been a reasonable conclusion. However, the agreement of the simple beam calculations and beam finite element model results with the experimental results may cast doubt on the accuracy some particular aspect of the solid-element model.

- There are not enough elements through the thickness in the solid-element model to allow for the bending stresses to be accurately calculated. While this is a possibility, closer examination of the maximum and minimum stresses predicted by the solid-element model shows that the neutral axis location (assuming a linear distribution of stress) is more than 0.5 in away from the centroid of the cross section. This result suggests that some other type of loading is being introduced into the beam.

Having considered and eliminated these possible explanations as likely, we are led to suspect the boundary conditions. The three-dimensional model made use of boundary conditions whose equivalent effect is to allow only rotation about the y-axis (along the knife-edge support). Indeed, the boundary condition illustrated in Fig. 4.7 seems to be a good representation of the physical constraint, as the real beam rests on a support that extends across the entire

Example 4.1: Four-Point Bend Test on a T-Beam (continued)

flange, and one is predisposed to visualizing this simple rotation condition. Note that the portion of the beam that extends beyond the support is not included in the finite element model.

However, the boundary conditions restrict displacements that are possible with the three-dimensional model and which *exceed* the conditions imposed by the actual knife-edge constraint. In particular, the flange of the beam does not remain perfectly flat. Rather, it rests freely on the support and is free to deform in the direction transverse to the beam's neutral axis and throughout the entire depth of the beam. Since the axial strain varies with distance away from the neutral axis, the transverse strain due to Poisson's ratio also varies. This variation of transverse strain, not accounted for in one-dimensional analyses, results in curvature of the flange. You can easily visualize this effect by bending a rubber eraser between thumb and forefinger and noticing the curvature transverse to the applied bending.

It nevertheless still seems reasonable that some three-dimensional model should work. We can modify the boundary conditions to allow the model to curve in the transverse direction. These alternative boundary conditions are *relaxed* to apply to the two corner nodes on each end of the beam only. The deflected shape of a slice of the beam section with these new boundary conditions applied is illustrated in Fig. 4.9. Although the deflections are greatly exaggerated, the tendency of the beam flange to curve rather than sit flat on the support is clearly evident. This relaxation of the constraint on the flange appears to have rather strong effects on the predicted bending stresses.

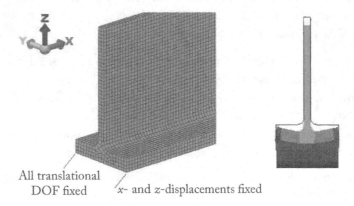

All translational DOF fixed   *x*- and *z*-displacements fixed

Figure 4.9: Modified boundary conditions applied to the finite element model result in an altered deformed shape of the beam at these supports.

Example 4.1: Four-Point Bend Test on a T-Beam (continued)

As reported in Table 4.3, the new results for peak bending stresses using the relaxed constraints are much closer to the experimental results than those using the stricter constraints.

Table 4.3: Results of alternate beam analyses

	$\sigma_{bottom}/\hat{\sigma}$	$\sigma_{top}/\hat{\sigma}$
Experiment	0.1108	-0.2464
Euler–Bernoulli beam theory	0.1134	-0.2724
FEA: beam elements	0.1134	-0.2724
FEA: solid elements, loosely-pinned supports	0.0946	-0.2230
FEA: solid elements, fully-pinned supports	0.0536	-0.2184

There are several important lessons to take away from this exercise.

1. With the loosely pinned supports, the error in maximum bending stress on the bottom of the beam is reduced from 52% to 16%, while the maximum bending stress on the top of the web is now even more accurate than the one-dimensional results.

2. For beams whose depth-to-span ratio is not small, Poisson effects on stresses may be significant. Furthermore, these effects are accentuated because the end constraints are placed along the beam flange surface which is not on the neutral axis. Beam theory inherently assumes that all constraints are placed at the neutral axis.

3. The one-dimensional results may agree well with experiment because of the proximity of the flange, where actual boundary conditions are placed in the experiment, to the actual neutral axis.

4. While a three-dimensional model can account for out-of-plane effects, the precise form of the boundary conditions can have strong effects on stresses.

5. Solid elements are not always the best choice for an analysis when this choice is made irrespective of the boundary conditions. Often, realistic deformations result that may be outside of the realm of one's limited experience. With easy access to a part or assembly modeled with a solid modeling program, it may seem logical to import and analyze the structure with three-dimensional elements for no more important reason than ease for the analyst.

Example 4.1: Four-Point Bend Test on a T-Beam (continued)

6. Very often the part or assembly modeled with a solid modeling program has been created without previous knowledge of where and how loads and boundary conditions will need to be applied in a subsequent finite element analysis. Often, analysts will struggle with wanting to import these solid model part or assembly files nonetheless. This may lead or even force them to place less than optimal loadings and boundary conditions where they otherwise might not.

7. Constraints that produce only negligibly small differences in strains can result in significant differences in internal stresses.

8. In this example problem, an analysis with over 14,000 three-dimensional solid elements produced *inferior* results compared to an analysis with four simple one-dimensional beam elements.

Use of three-dimensional analysis does not guarantee more accurate results. Because there are still discrepancies between the three-dimensional stress predictions and experimental results, we suggest that, as an exercise, students should further relax the boundary constraints along the flange to allow displacement along the beam's longitudinal axis. Since the flange is below the neutral axis, and there is bending, a compressive force will develop along the bottom of the flange if both ends of the beam are fixed in the axial direction. In this way, the extent to which this additional axial force does or does not affect the maximum bending stress can be explicitly determined.

SimCafe Tutorial 6: Large Depth-to-Span Ratio Beams

The purpose of this case study is to illustrate how assumptions of planar behavior affect numerical simulation of simple beam bending. The plane stress and plane strain assumptions lead to bounds on the actual three-dimensional behavior. While this analysis is simple to perform, the results are not so readily validated by those who are not ready to question when the planar approximations are reasonable to apply. Such reasoning can lead analysts to conclude that numerical results have "converged" on a result which is inaccurate by over 100%. The point of this exercise is to have analysts convince themselves that simplified theories are often bounds and that geometries that do not cleanly and unambiguously lend themselves obviously to either (thin or thick) limit, may still be ones for which one limit is reasonable and applicable. Also, an intuitive feel for making and applying these simplifications often still eludes users of the finite element method. This case study is an exercise in boundary condi-

SimCafe Tutorial 6: Large Depth-to-Span Ratio Beams (continued)

tion prescription, choosing appropriate finite element formulations, simulation convergence, and applying caution in interpreting one's results.

Follow the directions at `https://confluence.cornell.edu/display/ SIMULATION/2D+Beam` to complete the tutorial.

Example 4.2: Large Depth-to-Span Ratio Beams

A simply supported beam of rectangular cross section is point loaded at some arbitrary point along its length as shown in Fig. 4.10. Consider a beam where $L = 100\,$in, $a = 25\,$in, $h = 8\,$in, $b = 3\,$in, and load $P = 1000\,$lbf.

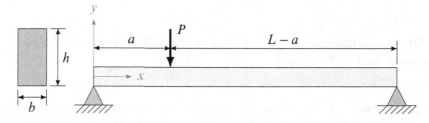

Figure 4.10: A beam with rectangular cross section is simply-supported while an off-center point load is applied.

While, in general, a finite element analysis will more accurately predict deflections than, say, internal stresses (we will discuss this in more detail in Chapter 5), this example illustrates a case in which even the deflections can be poorly modeled. We wish to examine the implications of analyzing the problem with one-, two-, and three-dimensional element formulations. Analysts must choose and defend their method of analysis including all implications that dimensional space imposes on the results. Examples of the simplest meshes for either plane stress or plane strain analyses using continuum elements are illustrated in Fig. 4.11.

Analysts may suppose that Euler-Bernoulli beam theory applies for these long, slender beams, presumably because it is the theory with which they are most familiar, and/or because it seems to work in other apparently similar examples, such as our previous example with the T-beam. However, while this assumption is intuitively appealing, we will see that it leads to a variety of pitfalls. Perhaps other expectation failures are in the offing.

Let us proceed with a narrative of this example assuming that the Euler-Bernoulli theory holds, although this has not yet been verified. Under this assumption, the curious analyst, in light of the previous example, might simulate the beam with several models. Due

Example 4.2: Large Depth-to-Span Ratio Beams (continued)

to the rectangular section of the beam, one-, two-, and three-dimensional models might be appropriate.

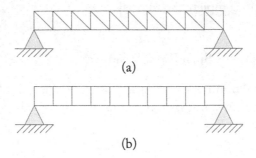

(a)

(b)

Figure 4.11: Typical mesh discretizations using (a) linear 3-node triangular elements and (b) bi-linear 4-node quadrilateral elements.

The analyst proceeds to simulate the beam using a variety of elements: one-dimensional beam elements, plane strain triangles, plane strain quadrilaterals, plane stress triangles, plane stress quadrilaterals, and three-dimensional brick elements (using what the analyst believes to be sufficiently relaxed end constraints, as per the previous example). The results for maximum deflection are reported in Fig. 4.12. All results are reported in dimensionless form, normalized by the characteristic deflection

$$\hat{v} = \frac{PL^3}{EI}.$$

According to these results, and still believing that Euler-Bernoulli beam theory is correct, the analyst would see that the maximum converged transverse deflection predicted by plane stress conditions underestimates the deflection predicted by Euler-Bernoulli beam theory by nearly 50%; by comparison, the maximum converged transverse deflection predicted by plane strain conditions overestimate the prediction of Euler-Bernoulli theory by 40%. The analyst also realizes that the converged results from the three-dimensional brick elements appear to be in agreement with the converged plane stress results, but that a coarse mesh instance of the plane strain model seems to agree well with the expected Euler-Bernoulli beam theory. How does the analyst sort out these mixed messages?

Example 4.2: Large Depth-to-Span Ratio Beams (continued)

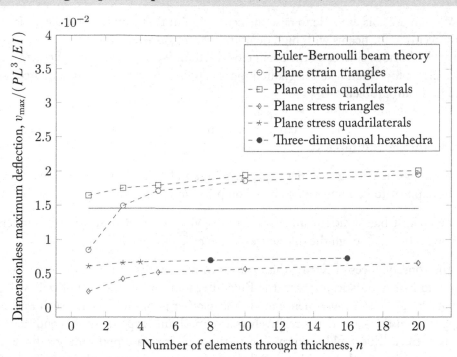

Figure 4.12: Maximum deflection predicted by finite element models assuming two-dimensional plane strain, two-dimensional plane stress, three-dimensional, and idealized one-dimensional behavior are compared with predictions from Euler-Bernoulli beam theory.

There is now a subtle point to make about this problem in comparison to the previous problem with the T-beam. Whereas in the previous problem the neutral axis was near the bottom of the beam, in this case it is not. Rather, it lies at mid-depth, and is hence far away from the location of the support pins that are at the bottom of the beam. This gives our first clue that regular Euler-Bernoulli beam theory is not applicable here.

Secondly, given that a fully three-dimensional analysis using solid elements can properly be specified to match the given boundary conditions, and given further that the three-dimensional formulation can account for the Poisson effects and out-of-plane curvature (which turn out to be significant), the three-dimensional analysis appears to give an accurate result. Because the converged plane stress solution agrees with the three-dimensional theory, and because the plane stress elements do not preclude out-of-plane Poisson effects, we have even further indication that the three-dimensional (and hence two-dimensional plane stress) solutions are valid.

Example 4.2: Large Depth-to-Span Ratio Beams (continued)

We can use this example to raise the general point that simple beam theory is not sufficiently accurate for beams with high depth to span ratios where the pinned boundaries are placed at the bottom surface of the beam cross section. Nevertheless, too often even experienced analysts too often take for granted that it applies universally as an accepted solution for slender beam problems.

We point out that this type of qualitative reasoning is not trivial. The analyst's ability to undertake this reasoning correctly depends on two key issues that have been articulated repeatedly throughout this text:

- the analyst is willing to anticipate, confront, and let go of misconceptions, even when they appear to be intuitive and based on prior understandings; and

- the analyst has sufficient understanding of Mechanics of Materials and understands how to think through the differences in the models considered.

The consequences of getting this analysis wrong, in this case, can be far reaching. The analyst who insists on sticking with the Euler-Bernoulli beam theory not only will persist with that error, but as a consequence might make other poor judgements, such as believing, as is apparent in this case, that a relatively coarse mesh under plane strain conditions is also generally correct! This could, in turn, lead to the analyst to not performing sufficient mesh refinement studies in other problems, and to accept other erroneous plane strain solutions.

In closing this example, we note that elementary beam theory would, in fact, be reasonable if one were to take in account the nature of the support boundary conditions. Some users will notice that the plane stress solutions converge to a maximum deflection nearly half that obtained by simple beam theory. In this case, they may investigate the possibility of pinning the end supports of the two-dimensional mesh at the mid-plane location of the neutral axis. This, of course, lowers the area moment of inertia by close to a factor of two, bringing the theory and two-dimensional analysis into very good agreement. Alternatively, they can apply beam theory employing offset neutral axes, i. e., one-dimensional beam element line models with a moment of inertia about some point well below the neutral axis as will be the case for pin supports on the bottom edge of the beam. This exercise illustrates the rather strong dependence of the solution of the boundary value problem on the precise prescription of the support boundary conditions, as well as the bounding nature of two-dimensional continuum approximations for truly three-dimensional problems.

When one accepts that the three-dimensional analysis is accurate, users can become understandably frustrated that a two-dimensional analysis is always an approximation whose accuracy they have to be prepared to verify. While a three-dimensional analysis may be accurate for this particular problem, it requires substantially more computational effort and cost than the corresponding two-dimensional plane stress approximation.

4.5   LESSONS LEARNED

These two case studies point out several realities in application of the finite element method.

1. When one proceeds to higher dimensions, while Poisson effects, i. e., lateral dimensional changes and out-of-plane warping, are captured, the precise manner in which classical boundary conditions such as simple supports or clamped supports are applied can have significant influence on the numerical results.

2. Improper boundary conditions can lead one to purposefully choose poorer element formulations and coarser meshes in attempts to validate a solution.

3. While one- and two-dimensional idealizations help reduce computational effort, they must be understood and substantiated.

These lessons illustrate several of the common errors encountered in using the finite element method [Chalice Engineering, LLC, 2009]. These include:

1. using wrong elements for an analysis,

2. incorrectly prescribing boundary conditions,

3. incorrectly applying theory for solution validation,

4. assuming finite element analysis is conservative, and

5. using finite element analysis for the sake of it.

There are arguably only two types of errors made in numerical simulation: either in faulty assumptions regarding the relevant physics governing the engineering system or discretization error in the numerical solution algorithm employed. Good analysts must understand and take responsibility for both. Modeling is, therefore, necessarily an iterative enterprise involving reassessing the validity of one's physical assumptions as one hones in on an acceptable solution. Because our numerical simulations are only approximations, this book has emphasized that users should be skeptical of their solutions prior to validating them. Further interesting reading regarding modeling approximation and anomalies can be found in Deaton [2010, 2013], Dvorak [2003], Fleenor [2009], Grieve [2006], and Kurowski [2001, 2002a,b,c].

CHAPTER 5

Wisdom Is Doing It

In theory, theory and practice are the same. In practice, they are not.

Albert Einstein

Do you know the difference between knowledge and wisdom? Wisdom is doing it!

Dan Millman
A Peaceful Warrior

Sometimes it is said that the application of science or a theory is "as much an art as a science." The practice of the finite element method fits the bill. Several authors have collected their own practical tips for application of the method. But, in general, books primarily about finite element theory do not present details regarding use of the method in practice. Books that attempt to address practical advice about applying the method in practice [Budynas, 2011, Kim and Sankar, 2009] almost always address issues that can be traced to the original list of ten most common mistakes presented in Chapter 1. Consider that the method is comprised of the following.

1. Preliminary analysis, which may entail:

    (a) simplifying the problem to obtain an analytical solution or estimation based on theory,

    (b) obtaining theoretical solutions representing upper or lower bounds for the solution, or

    (c) calculating the order of expected values for deflections and stresses and locations for their respective maxima/minima.

2. Pre-processing, which usually includes:

    (a) choosing an appropriate finite element formulation,

    (b) discretizing the domain globally,

    (c) refining it locally in areas of interest, and

    (d) applying loads and displacement constraints.

3. Solving the equations.

4. Post-processing the solution variables to compute

   (a) reaction forces and

   (b) internal stresses.

5. Interpreting and validating numerical solution results.

Referring to the list of most commonly made mistakes reported in Chapter 1, we attempt to correlate this list with the steps performed in the finite element method in Table 5.1. Five of the ten common errors might be avoided by paying particular attention to a well-performed preliminary analysis. Errors in pre-processing result in four of the typical errors. There is substantial overlap as preliminary analysis directly affects the most substantial step in pre-processing, which is discretizing the domain. Finally, three commonly made mistakes can be avoided with prudent post-processing. The solution of the equations for nodal point equilibrium usually results in no errors.

Note To The Instructor

While it is always important for students to know what a piece of computational software is doing on their behalf, having students mathematically carry out the steps of computing element equations, assembling them into a global matrix equation, reducing its rank once the boundary conditions have been decided, and solving the reduced set of equations will all be done for them in practice by commercial software. Because of the relative importance of the other mistakes they will likely make, we question the utility of assigning students problems requiring this mathematics. Many times, these are precisely the types of assignments that are given in an introductory course in the finite element method. It may behoove all of us who teach the method to realize that if we only have a single chance to speak to students on behalf of the method, we should at least discuss the list of places they will likely make mistakes. We also might be of better service to their education by assigning open-ended problems that require them to focus more on the steps where they are most likely to err while we are available to intervene and correct any ongoing misconceptions and poor practices before they become matters of routine.

Table 5.1: Mistakes listed by Chalice Engineering, LLC [2009] fall solely within portions of the analysis process performed by the analyst

Analysis Step	Mistakes Made
Preliminary Analysis	1,2,3,7,9
Pre-processing	3,6,9,10
Solution	0
Post-processing	2,4,5

5.1 PRELIMINARY ANALYSIS

Often, engineers go wrong early by ignoring what may arguably be the most important step. This is the preliminary analysis. This preliminary analysis takes place before one ever turns on the computer. Preliminary analysis consists of asking the question "What does one expect to happen?" To answer this question, one must apply mechanics theory. While most practical problems preclude analytical solutions, one can often simplify the problem to the extent where the order of deflection and stresses can be estimated and one can identify where their respective maxima are likely to occur. Sometimes simplifications of the real problem will lead to simpler solutions that may represent upper and lower bounds on deflections and stresses. Example 4.2 is a case in point. The finite element model is then created and analyzed to obtain a more precise, albeit approximate, solution whose quantitative results can be used for design purposes. When engineers neglect this step, they place themselves at a distinct disadvantage when attempting to later validate their numerical solution. It also places an analyst at a distinct disadvantage for pre-processing intelligently. It is often claimed by students that if they could compute the analytical solution, they wouldn't need the finite element method. But, in the end, this is a convenient rationalization to avoid the work involved in preliminary analysis. It is a crucial step if for no other reason than that it feeds so heavily into the most important decision made in pre-processing: discretizing the domain.

5.2 PRE-PROCESSING

Apart from preliminary analysis, the most common errors are made in pre-processing or establishing the numerical model of a real physical process. But preliminary analysis plays a critical role in reasons analysts go wrong in creating their models. Recall our discussion in Chapter 4 regarding what we are approximating, specifically the discussion centered on Figs. 4.2 and 4.3 for piecewise interpolation. Because there is no single, unique way to discretize the domain, creating a good quality mesh is a skill often best acquired through experience. Creating a good domain discretization requires first knowing something about the solution you are trying to approximate over that domain. This is because the finite element method approximates this solution with piecewise lower-order polynomial interpolations (the finite elements themselves). For instance, if one is trying to approximate a periodic solution using elements with linear interpolation, one should be asking the question "How many linear segments are required to sufficiently model a sinusoidal function over the domain prescribed?" So the decision of the element type, i. e., the solution interpolation polynomial order, and the decision on mesh density are intimately tied together given one knows something about the expected solution.

An equally important consideration is that the internal stresses in a deformed model are related to strains, which are higher-order derivatives of the displacement field. This has consequences that may best be illustrated by example. Consider a long, slender beam that is simply supported at both ends and loaded uniformly along its length as shown in Fig. 5.1. Over the entire domain, the exact bending moment varies quadratically and the exact shear force varies linearly.

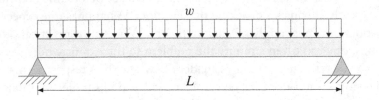

Figure 5.1: A uniformly loaded, simply supported, long, slender beam exhibits transverse deflection that varies as a fourth-order polynomial along its span.

The exact solutions are

$$v_{\text{exact}} = \frac{wL^4}{24EI} \left(-s + 2s^3 - s^4\right)$$
$$M_{\text{exact}} = \frac{wL^2}{2} \left(s - s^2\right)$$
$$V_{\text{exact}} = \frac{wL}{2} \left(1 - 2s\right),$$

where $s = x/L$. If we decide to model this beam with a mesh containing three cubic, one-dimensional beam elements, we will effectively be choosing to model the quartic function with three piecewise cubic functions. Consider a beam with length $L = 1$ ft uniformly loaded with $w = 24\,^{\text{lbf}}/_{\text{ft}}$, and $EI = 1\,\text{lbf} \cdot \text{ft}^2$. The normalized deflections predicted by this finite element model are shown in Fig. 5.2, which clearly have excellent agreement with the analytical solution.

Because bending moment varies linearly in a cubic beam element, the finite element prediction for the bending moment (and therefore the bending stress) over the beam then models a quadratic function with three linear segments. Finally, the linear variation of shear force is approximated by three piecewise constant segments. These FEA solutions are shown in Figs. 5.3 and 5.4, respectively. It is important to realize that while deflections are approximated well in this case, bending stress and shear force will only be captured reasonably well by successively localized refinements in mesh discretization.

Considering the correlation of the FEA prediction with the corresponding exact solution, the higher the order of the derivative of the displacement one wishes to approximate, the poorer the method does with any given mesh. The implication is that one generally needs a finer discretization to capture stresses accurately than to capture the deformed shape. In other words:

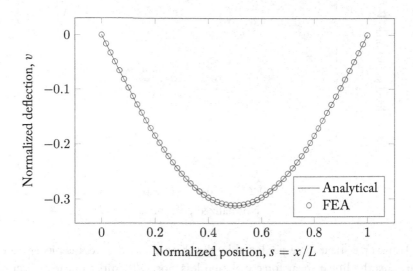

Figure 5.2: Displacement predicted by three piecewise continuous cubic interpolations very closely approximates the single quartic analytical variation in deflection.

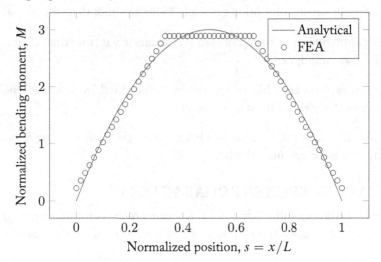

Figure 5.3: Bending moment predicted by the three element model is piecewise linear. This prediction captures the quadratic moment variation less closely than the cubic displacement interpolation captures the deformed shape.

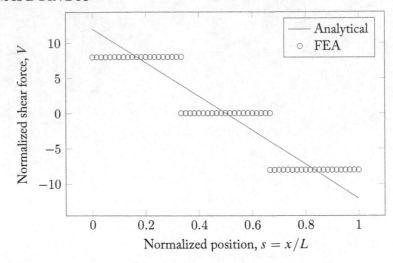

Figure 5.4: Transverse shear force predicted by the three element model is piecewise constant. This prediction captures the linear shear force variation less closely than the linear moment interpolation captures the quadratic bending moment.

1. the global displacement solution is generally more accurate than the global stress solution;

2. the discretization necessary to capture stresses accurately is finer than that needed to capture deformations accurately; and

3. what constitutes an acceptable mesh will be determined by whether one wishes a more accurate answer for deformation or stress.

It is, therefore, absolutely essential to know what element type one is using to properly mesh the problem domain and interpret one's results.

5.2.1 THE CAST OF ELEMENT CHARACTERS

An excellent presentation and discussion of practical element formulations is given in Budynas [2011]. Basically, one can place the majority of finite element formulations in one of five categories.

1. One-dimensional formulations for purely axial response (bar elements). Most elements are two-noded and utilize linear interpolation functions.

2. One-dimensional formulations that account for axial and out-of-plane bending response (beam elements). Most elements are two-noded and utilize cubic polynomial interpolation functions for transverse deflections.

3. Two-dimensional solid elements that account for two-dimensional in-plane stress states (plane stress/plane strain/axisymmetric solid or continuum elements). These elements may be triangular or quadrilateral. Typically, linear and parabolic interpolation functions are available. These effectively behave like two-dimensional analogs to one-dimensional bar elements.

4. Two-dimensional elements that respond to out-of-plane loads and moments (plate or shell elements). Plate elements effectively behave like two-dimensional analogs to one-dimensional beam elements.

5. Full three-dimensional solid elements. Typical elements are tetrahedral and hexahedral (brick) elements. Both linear and parabolic interpolation functions are offered in most element libraries.

These element formulations are illustrated in Table 5.2.
With regard to specific element formulations:

1. One-dimensional element formulations cannot capture stress concentrations and should be avoided where such stress risers are expected.

2. Two-dimensional element formulations can reduce the computational mesh size by orders of magnitude when conditions of plane stress, plane strain, or axisymmetry apply. Two-dimensional analysis should be considered in these limits.

3. For such two-dimensional element formulations, generally quadrilateral elements (of the same order interpolation) outperform triangular elements.

4. Two-dimensional element formulations used to capture in-plane bending should contain a minimum of three to five elements across the cross section perpendicular to the bending axis. Generally, where in-plane bending occurs in two-dimensional analysis, one should consider use of a higher-order, usually parabolic element interpolation.

5. One should avoid using three-noded triangular plate elements for out-of-plane bending as they are particularly stiff. In such analyses, the number of degrees of freedom necessary for a convergent solution will often dictate use of a higher-order element interpolation.

6. For three-dimensional solid analysis, hexahedral (brick) elements generally outperform tetrahedral elements, but tetrahedral elements will often be used by automated mesh generators because they can most easily fill generally complex three-dimensional regions.

7. When using tetrahedral element formulations in three-dimensional analysis, it is preferred to use a higher order, i. e., parabolic interpolation of displacements.

8. Three-dimensional solid elements often do not include rotational nodal degrees of freedom. Therefore, modeling global rotation at a boundary becomes yet a further approximation.

Table 5.2: Basic finite element types

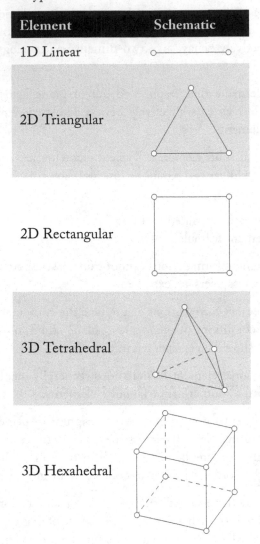

Element	Schematic
1D Linear	
2D Triangular	
2D Rectangular	
3D Tetrahedral	
3D Hexahedral	

5.2.2   GOOD AND BAD ELEMENTS

Good quality meshes typically employ:

- aspect ratios as close to unity as is feasible, i. e., equal side lengths in any single element;

- element shapes that avoid irregularities such as excessively small or large corner (skew) angles, e. g., 90° angles in quadrilaterals and 60° angles in triangles;

- gradual transition in element size. Rapid transitions in element size should be avoided whenever possible; and

- mesh refinement where the stress gradients are large.

Poor quality elements will inevitably appear in complex geometries, particularly when an analyst employs automatic mesh generators. Typically, commercial software will flag such elements with warnings and alert the user. An analyst is then responsible for adjusting the mesh locally, perhaps manually if necessary, to assure good quality results.

In general, it is difficult to avoid having an arbitrarily oriented element in a region of constant stress in a mesh. For this reason, element formulations are tested to ensure they can predict reasonably constant stress values in such cases. This is called the patch test. All good elements should be able to pass the patch test [Irons and Shrive, 1983].

A nice discussion of good and bad element behaviors is presented in Irons and Shrive [1983] and Kim and Sankar [2009]. Many good meshing strategies are outlined by Budynas [2011].

5.2.3 APPLYING BOUNDARY CONSTRAINTS

Several rules of thumb are necessary to consider when applying boundary constraints.

1. Be sure the boundary conditions applied to the model always remove all rigid body translation and rotation, i. e., "always tie down the horse." Some commercial software packages will attempt to solve such ill-posed problems and deliver no results.

2. Errors in boundary conditions can be subtle and hard to recognize. For example, consider the two-dimensional constraints applied for the simple supports in Example 4.2. Because they are not applied along the neutral axis of the beam, the apparent flexural stiffness of the beam is nearly twice that one would calculate using elementary beam theory.

3. Applying idealized boundary conditions becomes more difficult in higher dimensions. For instance, applying a simple support is straightforward in one-dimensional elements, but in two and three dimensions, there are multiple ways to apply the constraint at the domain edges. This same quandary occurs when applying any idealized boundary constraint such as a clamped edge in two or three dimensions where rotational degrees of freedom are not available and constraints on the local slope of the deformed structure cannot be explicitly constrained.

4. Local results, particularly maximum deflections or stresses, can be very sensitive to small variations in the application of boundary condition constraints.

5.2.4 APPLYING EXTERNAL LOADS

Several rules of thumb should be considered when applying loads to the structure.

1. Point loads are idealized load applications and will generally result in unreasonably large internal stresses in the vicinity of the application point. One should consider applying localized pressures when possible.

2. Usually, when concentrated loads are applied the stresses resulting from statically equivalent loads will be independent of the method of application a distance away from the load that is of the order of the transverse dimensions of the structure locally. This is the principle of St. Venant. It should be employed liberally in application of the finite element method and in interpreting its results.

3. In an analogous manner as with prescribing boundary constraints, when one models domains in two or three dimensions, element formulations may not have rotational degrees of freedom. For such cases, application of a concentrated moment or couple is no longer unique and not as straightforward as it is when using one-dimensional elements. In such cases, one should consider experimenting with different possible prescriptions of the couple using local point loads and compare stresses a St. Venant's decay distance away from the concentrated moment.

4. Generally, the order of complexity of the solution to boundary value problems will increase with the order of the loading. Given a specific finite element formulation, the more complex the loading, the more approximate the solution. This was illustrated in the beam example of Fig. 5.1. Such one-dimensional beam elements capture bending stress and shear forces exactly when only point loads and couples are applied. These same bending stresses and shear forces are only approximately predicted when distributed or more complex loading is applied.

5.3   POST-PROCESSING

> Computer graphics has achieved such a level of polish and versatility as to inspire great trust in the underlying analysis, a trust that may be unwarranted. (One can now make mistakes with more confidence than ever before.)
>
> R.D. Cook, D.S. Malkus, and M.E. Plesha
> Concepts and Applications of Finite Element Analysis,
> 3rd Edition

There are several rules of thumb to consider when post-processing results.

1. Plotting deformed shapes of structures is a good way to spot particular errors in application of boundary constraints.

2. Element stresses are most accurate at internal integration points where they are calculated. These stresses are averaged at nodes shared by elements. The nodal-averaged stresses are

interpolated between nodes, contoured, and then, generally, artificially smoothed to create contoured results.

3. When displaying stress contours, it is often good practice to contour element values directly as well as the nodally averaged values. This is a good practice because:

   (a) If the element stresses are observably discontinuous to the eye, then the stress gradients are larger than the mesh is capable of predicting and one should refine the mesh.

   (b) If the element stresses are not overly discontinuous, then the smoothed contours are sufficient to represent the overall character of the solution.

5.4   FURTHER RULES TO LIVE BY IN PRACTICE

One can establish a set of ground rules that can serve as a starting point for good practical finite element analysis. Again, this list, while not exhaustive, attempts to address several of the most common errors made in applying the finite element method.

1. Use the finite element method only when it is necessary, i. e., when the simplest formulae outlined in Chapter 2 or other analytical methods are not generally applicable.

2. There are no units involved in formulation of the finite element method. An analyst must always use dimensionally consistent units and interpret results accordingly.

3. The finite element discretization results in a model that is *too stiff*, implying:

   (a) models upon which only displacement boundary conditions are applied will, in general, result in stresses that are higher than the actual stresses;

   (b) models upon which only force boundary conditions are applied will, in general, result in displacements that are smaller than the actual displacements; and

   (c) no general conclusions can be made once the boundary constraints are mixed, which is most often the case.

4. One should not generally assume that finite element analysis is conservative.

5. It is not necessarily true that three-dimensional analysis outperforms two-dimensional analysis or that two-dimensional analysis outperforms one-dimensional analysis.

6. One should consider mesh refinements in regions where there are large gradients in material stiffness such as dissimilar material interfaces or large discontinuities in load-bearing areas.

7. Consider applying the principle of St. Venant in order to avoid modeling geometric features wherein the stress results are not of primary importance, e. g., details at or near load application points.

8. Exploit global symmetry wherever and as much as possible.

9. When importing geometries from solid modeling software, it is important, when possible, to create the solid model with design intent. By this, we mean that solid geometry entities such as grid points and surfaces should be strategically created such that boundary conditions can be placed on nodes and element edges that lie, respectively, on these solid entities. This practice allows one to usually perform mesh refinements and iterations without the inconvenience of re-applying the boundary conditions.

5.5 SOLUTION VALIDATION

> Believe nothing, no matter where you read it or who said it, unless it agrees with your own reason and your own common sense.

Buddha

> Nobody believes a model except the one who devised it; everyone believes an experiment except the one who performed it.

Albert Einstein

Perhaps not all experimentalists are so cautious nor all modelers as careless, but, as evidenced by the common errors made by analysts, it can seem as if those who computationally model systems can be led to a false sense of security in their numerical solutions. We like to recommend that all numerical model results must, initially at least, be viewed through skeptical spectacles. If one treats at least one's initial findings as *guilty until proven innocent*, one will be less likely to accept results that are incorrect.

In general, an engineering analysis can be accomplished either

1. theoretically, from first principles,

2. approximately, using numerical analysis, or

3. empirically, using discrete experiments.

Having all three one might consider the mother lode. But, in any analysis, we should shoot for results of one approach to be benchmarked or validated by one or both of the others. Here we define validation as the process of determining the degree to which a model is an accurate representation of the real world from the perspective of the intended uses of the model. In essence, validation provides evidence that the correct model is solved to a given level of accuracy.

As we are attempting to prove the results of our numerical analyses innocent, we should validate all results with either theoretical results or experimental data. While theoretical results are often precluded in real applications, they may have limited applicability when they represent

1. upper and lower bounds of the real solution or

2. the correct solution in only part of the global domain.

When using experimental results for validation, one should consider the following.

1. They are often considered the harbinger of truth.

2. Boundary constraints more easily realized in the laboratory can sometimes be difficult to realize in a computational model, for example, machine compliance for a tensile test specimen.

3. Boundary constraints more easily realized in discrete analysis can sometimes be more difficult to achieve in the laboratory.

4. Experiments can be costly and time-consuming.

Numerical analyses should not be trusted without either theoretically or experimentally validating the solution. Neither should the results of numerical analyses be accepted without proper examination of insensitivity to the mesh discretization. As in Example 4.2, a proper convergence study should always be attempted. Correct results can only be obtained in the limit as the results are no longer sensitive to the use of any finer discretization of the global domain. We term such convergence *mesh insensitivity*. When the results fall within a specified insensitivity to the mesh or element size, one can conclude the numerical analysis has converged. It is important to note that this is a necessary but not sufficient condition for the computational results to be acceptable or a correct solution to the problem posed. Recall that the solutions in Example 4.2 eventually converged, but those assuming plane strain conditions were incorrect, i. e., they solved the wrong problem.

5.6   VERIFICATION

> Extensive tests showed that many software codes widely used in science and engineering are not as accurate as we would like to think.
>
> Les Hatton
> Oakwood Computing

By verification, we refer to the process of determining that a model implementation accurately represents the developer's conceptual description. Verification provides evidence that the numerical model is solved correctly. It is tacitly assumed that commercial software is completely debugged before a version is released. Les Hatton at Oakwood Computing has presented interesting findings that indicate errors in software and programming, while small in number, do occur. This sometimes happens in commercial FEA software. Most instances of which we are aware have been in the post-processing software. While the primary variable solution is most often entirely correct, sometimes listings and contour plot variables are not stored correctly and are subsequently improperly displayed. Luckily, these instances are rare and not a primary cause of errors on the part of the analyst. In any case, they can be caught by prudent use of preliminary analysis.

We believe that the majority of textbooks addressing introductory finite elements primarily and predominantly emphasize the mathematical foundation and procedural application of the method. We have emphasized, rather, a practical approach based on recognition that most errors made in application of the method are in pre- and post-processing and are made mostly in model development. Further interesting reading regarding issues of practical application of the finite element method can be found in Dunder and Ridlon [1978], Dvorak [2003], Gokhale et al. [2008], Morris [2008], and Sastry [2010].

Summary

The most common mistakes made by novice users of the finite element method involve procedural steps performed explicitly by the user. Exercises in many textbooks emphasize mathematical elements of the procedure performed strictly by the computer. We have introduced an alternative examination of the method used in practice that focuses on a published list of commonly made mistakes. Examination of the root causes of such mistakes reveals that they are intimately tied more to a user's command of underlying theory of strength of materials and less to a user's ability to reproduce mathematical computations undertaken by the processor.

We outlined a basic requisite skill set necessary to undertake use of the finite element method. Then we explored excursions where first the underlying theory no longer holds, and then ultimately where users are most likely to interface with the software in a faulty manner. Finally, we posited a short listing of rules for applying the finite element method in practice. While this list is generally acknowledged by many practitioners, we find that it is typically relegated to more of an aside and less of a central theme. We provided relatively simple examples to showcase where mistakes are made when one does not follow practical rules of thumb from the start.

If the method is taught with more of this emphasis on expectation failures of newly learned mechanics of materials, and more prudent attention to questioning computational complacency, it is our hope that the occurrence of these common mistakes may be reduced. Also, an earlier introduction to the method as a practical tool may prove to be a useful precursor to better and deeper learning of the mathematics underlying finite element interpolation. We argue that, instead of emphasizing steps performed well by the computer, becoming competent in finite element analysis should focus on the steps of the process where analyst's choices have the greatest impact on the results.

Afterword

This book was written to supplement texts on FEA theory with prudent rules for practice by focusing specifically on errors commonly made in industry. Based on our experience teaching the method to undergraduates, we included examples where students have faltered in the past and couched these in terms of expectation failures. After reading this book, if you have comments on the presentation of the exercises or wish to suggest additional examples that emphasize expectation failures, feel free to contact the authors at LBAcomments@gmail.com. Thank you, in advance, for any input you have.

Bibliography

Allain, R. (2011). *Just Enough Physics*. Amazon Digital Services. 27

Bain, K. (2004). *What the Best College Teachers Do*. Harvard University Press. xiii, 5, 48

Bhaskaran, R. (2012). SimCafe Wiki-Based Online Learning System. In `https://confluence.cornell.edu/display/SIMULATION/Home`. xiv

Bhaskaran, R. and Dimiduk, K. (2010). Integrating advanced simulations into engineering curricula: Helping students to approach simulation like experts. In *NSF Award 0942706*. xiv

Brooks, D. (2013). The Practical University. *The New York Times*, April 5:A23. xiii

Bruner, J. S. (1960). *The Process of Education*. Harvard University Press. xii

Budynas, R. G. (2011). *Advanced Strength and Applied Stress Analysis*. McGraw Hill. 69, 74, 77

Carson, E. and Cobelli, C. (2000). *Modelling methodology for physiology and medicine*. Academic Press. 50

Chalice Engineering, LLC (2009). Ten common mistakes made in finite element analysis. In `http://www.chalice-engineering.com/analysis_basics/Top_Ten_mistakes.html`. 3, 7, 67, 70

Conly, S. (2013). Three Cheers for the Nanny State. *New York Times*, March 24:A26. xv, 1, 5

Cook, R. D., Malkus, D. S., Plesha, M. E., and Witt, R. J. (2002). *Concepts and Applications in Finite Element Analysis, 4th Edition*. McGraw Hill. xiv

Deaton, B. (2010). Believing experiments vs. simulations. In `http://onlyamodel.com/2010/quote-experiments-vs-simulations/`. 67

Deaton, B. (2013). Responding to skepticism toward your model. In `http://onlyamodel.com/2013/responding-to-skepticism-toward-your-model/`. 67

du Toit, J., Gosz, M., and Sandberg, G. (2007). Minisymposium on the Teaching of Finite Elements at the Undergraduate Level. In *9th U.S. National Congress on Computational Mechanics*. 9

Dunder, V. F. and Ridlon, S. A. (1978). Practical applications of the finite element method. *Journal of the Structural Division ASCE*, 104:9–21. 82

Dvorak, P. (2003). A Few Best Practices for FEA Users. In `http://machinedesign.com/article/a-few-best-practices-for-fea-users-0904`. 67, 82

Dym, C. (2004). *Principles of Mathematical Modeling*. Academic Press, 2 edition. 49, 50

Fleenor, M. (2009). Modeling Stupidity. *The Teaching Professor*, 23:2–4. 67

Gieck, K. and Gieck, R. (2006). *Engineering Formulas*. Gieck Verlag Publishing. 45

Gokhale, N., Deshpande, S., Bedekar, S., and Thite, A. (2008). *Practical Finite Element Analysis*. Finite to Infinite Publishers. 82

Greenbaum, A. and Chartier, T. P. (2012). *Numerical Methods: Design, Analysis, and Computer Implementation Algorithms*. Princeton University Press, Princeton, New Jersey. 50

Grieve, D. J. (2006). Errors Arising in FEA. In `http://www.tech.plym.ac.uk/sme/mech335/feaerrors.htm`. 67

Hake, R. R. (1998). Interactive Engagement vs. Traditional Methods: A Six Thousand Student Survey of Mechanics Test Data for Introductory Physics Courses. *American Journal of Physics*, 66:64–74. DOI: 10.1119/1.18809. xii, 5

Hatton, L. (1999). Programming Technology, Reliability, Safety, and Measurement. In `www.leshatton.org/wp-content/uploads/2012/01/PTRel_IER298.pdf`. 9

Irons, B. and Shrive, N. (1983). *Finite Element Primer*. Ellis Horwood Publishers. 56, 77

Jeremić, B. (2009). Verification and Validation in Geomechanics. In *A Multidisciplinary Workshop on Deformation and Failure of Geomaterials*, Brindisi, Italy. 7, 45

Kim, N.-H. and Sankar, B. V. (2009). *Introduction to Finite Element Analysis and Design*. J Wiley and Sons. xiv, 69, 77

Kurowski, P. (2001). Easily Made Errors Mar FEA Results. In `http://machinedesign.com/article/easily-made-errors-mar-fea-results-0913`. 67

Kurowski, P. (2002a). How to Find Errors in Finite Element Models. In `http://machinedesign.com/article/how-to-find-errors-in-finite-element-models-1115`. 67

Kurowski, P. (2002b). More Errors that Mar FEA Results. In `http://machinedesign.com/article/more-errors-that-mar-fea-results-0321`. 67

Kurowski, P. (2002c). When Good Engineers Deliver Bad FEA. In `http://machinedesign.com/article/when-good-engineers-deliver-bad-fea-1115`. 67

Kurowski, P. (2013). *Engineering Analysis with SolidWorks Simulation 2013*. SDC Publications. xiv

Lawrence, K. L. (2012). *ANSYS Workbench Tutorial Release 14*. SDC Publications. xiv

Lee, H.-H. (2012). *Finite Element Simulations with ANSYS Workbench 14*. SDC Publications. xiv

Logan, D. L. (2001). *Applications in the Finite Element Method*. Brooks Cole Publishing Company. xiv

McDermott, L. C. (1984). Research on Conceptual Understanding in Mechanics. *Physics Today*, 37:24–34. DOI: 10.1063/1.2916318. xii, 5

McDermott, L. C. (2001). Oersted Medal Lecture 2001: Physics Education Research: The Key to Student Learning. *American Journal of Physics*, 69:1127–1137. DOI: 10.1119/1.1389280. xiii

Montfort, D., Brown, S., and Pollack, D. (2009). An Investigation of Students' Conceptual Understanding in Related Sophomore to Graduate-Level Engineering and Mechanics Courses. *Journal of Engineering Education*, 98:111–129. DOI: 10.1002/j.2168-9830.2009.tb01011.x. xii, 5

Morris, A. (2008). *A Practical Guide to Reliable Finite Element Modeling*. J. Wiley and Sons. DOI: 10.1002/9780470512111. 82

Papadopoulos, C. (2008). Assessing Cognitive Reasoning and Learning in Mechanics. In *American Society for Engineering Education Annual Conference and Exposition*. xii, 5, 33

Papadopoulos, C., Roman, A. S., Gauthier, G. P., and Ponce, A. (2013). Leveraging Simulation Tools to Deliver Ill-Structured Problems in Statics and Mechanics of Materials: Initial Results. In *American Society for Engineering Education Annual Conference and Exposition*. 22, 29

Papadopoulos, J., Papadopoulos, C., and Prantil, V. C. (2011). A Philosophy of Integrating FEA Practice Throughout the Undergraduate CE/ME Curriculum. In *American Society for Engineering Education Annual Conference and Exposition*. xii, 3, 8, 12, 33, 37, 45

Paulino, G. (2000). Warning: The Computed Answer May Be Wrong. In http://paulino. cee.illinois.edu/courses/cee361/handouts/wrcabm.htm. 1, 5, 9

Philpot, T. A. (2010). *Mechanics of Materials: An Integrated Learning System*. J. Wiley and Sons. 27, 33

Pope, J. E., editor (1997). *Rules of Thumb for Mechanical Engineers: A Manual for Quick, Accurate Solutions to Everyday Mechanical Engineering Problems*. ROTpub. 27

Prantil, V. C. and Howard, W. E. (2007). Teaching Finite Element Simulation in Conjunction with Experiment and Theory in an Integrated Systems Design. In *9th U.S. National Congress on Computational Mechanics*. 45

Prantil, V. C. and Howard, W. E. (2008). Incorporating Expectation Failures in an Undergraduate Finite Element Course. In *American Society for Engineering Education Annual Conference and Exposition*, volume 1. ASEE, Curran Associates, Inc. 45

Public Broadcasting System–NOVA (1993). The Best Mind Since Einstein - Richard Feynman Biography. Television Production. 49

Riley, W. F., Sturges, L. D., and Morris, D. H. (2007). *Mechanics of Materials, 6th Edition*. J Wiley and Sons. 33

Sastry, S. S. (2010). Accepted Practices in Practical Finite Element Analysis of Structures. In `http://www.nafems.org/downloads/india/webinar/mar_10/accepted_fe_practices_nafems_india.pdf`. 82

Solverson, R. (1953). Stress Concentrations in Fillets. Master's thesis, California Institute of Technology. 35

Steif, P. S. (2012). *Mechanics of Materials: An Integrated Learning System*. J Wiley and Sons. 27, 33

Streveler, R., Litzinger, T., Miller, R., and Steif, P. (2008). Learning Conceptual Knowledge in the Engineering Sciences: Overview and Future Research Directions. *Journal of Engineering Education*, 97:279–294. DOI: 10.1002/j.2168-9830.2008.tb00979.x. xii, 5

Thompson, E. G. (2004). *Introduction to the Finite Element Method: Theory, Programming and Applications*. John Wiley and Sons. xiv

Young, W. C. and Budynas, R. G. (2002). *Roark's Formulas for Stress and Strain, 7th Edition*. McGraw Hill. 35, 45

Zienkiewicz, O. and Taylor, R. (2005). *The Finite Element Method for Solid & Structural Mechanics, 6th Edition*. Elsevier, Butterworth-Heinemann Publishing. xiv

Zienkiewicz, O., Taylor, R., and Zhu, J. (2005). *The Finite Element Method: Its Basis & Fundamentals, 6th Edition*. Elsevier, Butterworth-Heinemann Publishing. xiv

Authors' Biographies

VINCENT C. PRANTIL

Vincent C. Prantil earned his B.S., M.S., and Ph.D. in Mechanical Engineering from Cornell University where he was awarded The Sibley Prize in Mechanical Engineering and held an Andrew Dickson White Presidential Fellowship. He was a Senior Member of Technical Staff at Sandia National Laboratories California in the Applied Mechanics and Materials Modeling Directorates for 11 years. His research interests lie in microstructural material modeling, dry granular materials, metals plasticity, finite element, and numerical analysis. He was jointly awarded an R&D100 award for co-developing the Sandia Microstructure-Property Model Software in 2000 and held the Otto Maha Research Fellowship in Fluid Power at the Milwaukee School of Engineering (MSOE) from 2006–2008. He joined the faculty in the Department of Mechanical Engineering at MSOE in September 2000 where he presently specializes in finite element model development, numerical methods, and dynamic systems modeling.

CHRISTOPHER PAPADOPOULOS

Christopher Papadopoulos earned B.S. degrees in Civil Engineering and Mathematics in 1993 at Carnegie Mellon University, and his Ph.D. in Theoretical and Applied Mechanics in 1999 at Cornell University, where he was a National Science Foundation Graduate Research Fellow. He is currently a member of the faculty of the Department of Engineering Science and Materials at the University of Puerto Rico, Mayagüez (UPRM), where he has worked since 2009. He was previously a member of the faculty in the Department of Civil Engineering and Mechanics at the University of Wisconsin–Milwaukee from 2001–2008. Chris is currently the principal investigator of two NSF projects, one in appropriate technology and engineering ethics, and the other in mechanics education. He has additional research interests in nonlinear structural mechanics and biomechanics. Chris currently serves as Secretary and Executive Board Member of the ASEE Mechanics Division and he is the chair of the Mechanics Committee in his department. He is also a member of a campus committee that arranged for an art exhibit honoring the life of Roberto Clemente to be donated to the UPRM campus from the Smithsonian Museum. Chris is a passionate educator and advocate for humanitarian uses of technology. In his free time he enjoys swimming, cycling, running, cooking, and learning the languages of the Caribbean.

PAUL D. GESSLER

Paul D. Gessler is currently a graduate student pursuing his M.S. in the Mechanical Engineering Department at Marquette University in Milwaukee, Wisconsin. He earned his B.S. in Mechanical Engineering from the Milwaukee School of Engineering in 2012. Paul's main interests are using modeling and simulation at an appropriate abstraction level to improve the product design and systems engineering process. He has experience with a wide variety of commercial FEA/CFD codes and has written several bespoke codes for fluid, structural, and thermal system analysis. Paul hopes to be a proponent of model-based design practices in industry throughout his career.

Index

Printed in the United States
by Baker & Taylor Publisher Services